AF701670

Exploring Electric Kiln Techniques

Related Titles published by The American Ceramic Society

Pioneer Pottery
Co-published by The American Ceramic Society, Westerville, Ohio, USA and A&C Black, London, England
Michael Cardew
©2002, ISBN 1-57498-125-0

Raku
Co-published by The American Ceramic Society, Westerville, Ohio, USA and A&C Black, London, England
John Mathieson
©2002, ISBN 1-57498-166-8

Rudy Autio
Louana Lackey
©2002, ISBN 1-57498-144-7

Clay: A Studio Handbook
Vince Pitelka
©2001, ISBN 1-57498-090-4

Mary Wondrausch on Slipware
Co-published by The American Ceramic Society, Westerville, Ohio, USA and A&C Black, London, England
Mary Wondrausch
©2001, ISBN 1-57498-149-8

Wood Firing: Journeys and Techniques
With a Foreword by Dick Lehman
©2001, ISBN 1-57498-143-9

Creative Ideas for Clay Artists
Edited by Anderson Turner
©2001, ISBN 1-57498-122-6

Barrel, Pit, and Saggar Firing
Edited by Sumi von Dassow
©2001, ISBN 1-57498-127-7

Out of the Earth Into the Fire: A Course in Ceramic Materials for the Studio Potter, Second Edition
Mimi Obstler
©2000, ISBN 1-57498-078-5

The Extruder Book
Daryl E. Baird
©2000, ISBN 1-57498-073-4

The Potter's Professional Handbook
Steven Branfman
©1999, ISBN 1-57498-173-0

Setting up a Pottery Workshop
Co-published by The American Ceramic Society, Westerville, Ohio, USA and A&C Black, London, England
Alistair Young
©1999, ISBN 1-57498-106-4

Glazes for the Craft Potter, Revised Edition
Co-published by The American Ceramic Society, Westerville, Ohio, USA and A&C Black, London, England
Harry Fraser
©1998, ISBN 1-57498-076-9

Answers to Potters' Questions II
Edited by Ruth C. Butler
©1998, ISBN 1-57498-085-8

Great Ideas for Potters II
Edited by Ruth C. Butler
©1998, ISBN 1-57498-068-8

Answers to Potters' Questions
Edited by Barbara Tipton
©1990, ISBN 0-934706-10-7

Great Ideas for Potters
Edited by Barbara Tipton
©1983, ISBN 0-934706-09-3

Potter's Wheel Projects
Edited by Thomas Sellers
©1968, ISBN 0-934706-04-2

Decorating Pottery
F. Carlton Ball
©1967, ISBN 0-934706-05-0

Brush Decoration for Ceramics
Marc Bellaire
©1964, ISBN 0-934706-02-6

Ceramic Projects
Edited by Thomas Sellers
©1963, ISBN 0-934706-08-5

Throwing on the Potter's Wheel
Thomas Sellers
©1960, ISBN 0-934706-03-4

Underglaze Decoration
Marc Bellaire
©1957, ISBN 0-934706-01-8

Copper Enameling
Jo Rebert and Jean O'Hara
©1956, ISBN 0-934706-00-X

For information on ordering titles published by The American Ceramic Society, or request a ceramic art publications catalog, please contact Customer Service Department at 614-794-5890 (phone), 614-794-5892 (fax), customersrvc@acers.org (e-mail), or write to Customer Service Department, 735 Ceramic Place, Westerville, OH 43081, USA.

Subscribe to Clayart!
Clayart is the "electronic voice of potters worldwide," sponsored by The American Ceramic Society. Subscriber-initiated discussions range from questions/answers on materials and techniques to business advice and philosophical debate. Visit the website and subscribe at www.ceramics.org/clayart.

Visit our online bookstore at www.ceramics.org.

Exploring Electric Kiln Techniques

A Collection of Articles from *Ceramics Monthly*

Edited by Sumi von Dassow

Published by

The American Ceramic Society
735 Ceramic Place
Westerville, Ohio 43081 USA

The American Ceramic Society

Founded in 1953, *Ceramics Monthly* is an internationally distributed magazine on ceramic art and craft. Each issue includes up-to-date information on exhibitions, available materials and trends, as well as profiles of individual artists, critical analyses, production processes, and clay and glaze recipes. While principally covering contemporary work, the magazine also looks back at influential artists and events from the past.

The American Ceramic Society
735 Ceramic Place
Westerville, Ohio 43081

©2002 by The American Ceramic Society.
All rights reserved. Published 2001.
Printed in the United States of America.

05 04 03 02 01 5 4 3 2 1

ISBN: 1-57498-160-9

Cover design by Melissa Bury, Columbus, Ohio and David Houghton, ACerS
Cover image: "Forbidden Flower" by Keisuke Mizuno.

No part of this book may be reproduced, stored in a retrieval system, or transmitted in any form or by any means, electronic, mechanical, photocopying, microfilming, recording, or otherwise, without the written permission from the publisher, except by a reviewer, who may quote brief passages in review.

Authorization to photocopy for internal or personal use beyond the limits of Sections 107 and 108 of the U.S. Copyright Law is granted by The American Ceramic Society, provided that the appropriate fee is paid directly to the Copyright Clearance Center, Inc., 222 Rosewood Drive, Danvers, MA 01923 USA, www.copyright.com. Prior to photocopying items for educational classroom use, please contact Copyright Clearance Center, Inc. This consent does not extend to copyright items for general distribution or for advertising or promotional purposes or to republishing items in whole or in part in any work in any format. Requests for special photocopying permission and reprint requests should be directed to Senior Director, Publications, The American Ceramic Society, 735 Ceramic Place, Westerville OH 43081 USA.

Statements of fact and opinion are the responsibility of the authors alone and do not imply an opinion on the part of the officers, staff, or members of The American Ceramic Society. The American Ceramic Society assumes no responsibility for the statements and opinions advanced by the contributors to its publications or by the speakers at its programs. Registered names and trademarks, etc., used in this publication, even without specific indication thereof, are not to be considered unprotected by the law.

For more information on ordering books published by The American Ceramic Society, subscribing to our publications—including *Ceramics Monthly*— or to request a publications catalog, please call 614-794-5890 or visit our online bookstore at www.ceramics.org.

CONTENTS

4 MAJOLICA—FROM THE RENAISSANCE TO THE SPACE AGE

The electric kiln breathes new life into an age-old decorating technique.

5 THE BUILDING BLOCKS OF ELECTRIC KILN POTTERY—CLAY AND GLAZE FORMULATION

Recipes and research ideas for electric kiln clays and glazes.

INTRODUCTION

Power to the People

The latter part of the twentieth century saw an explosion in the number of people doing pottery. Art schools, community art and recreation centers, and senior centers offer pottery classes for all ages at affordable rates, and a large number of these students go on to eventually set up home studios. This popularity has largely been a result of the development of the electric kiln which, compared to gas or wood-burning kilns, is small and portable, inexpensive, and almost as easy to purchase, install and operate as a refrigerator or a washing machine. The electric kiln has, very literally, brought power to the people and made it possible for anyone to try their hands at pottery.

Of course, this did not happen overnight with the introduction of the electric kiln. Since its invention, potters have been exploring its capabilities, eager to discover just what it can do that is unique and exciting. From the point of view of the pottery inside the kiln, the difference between an electric kiln and a gas or wood kiln is the simple fact that there is no fuel burning in an electric kiln, thus no demand for oxygen to keep the fire going. This difference can have a profound effect on the glazes in the kiln. Many of the treasured glazes of the Orient—temmoku and oxblood and celedon, valued for their beauty and simplicity—rely on a reduction, or oxygen-poor, atmosphere in the kiln for their effect and are difficult or impossible to duplicate in the neutral, or oxidation atmosphere of the electric kiln. On the other hand, many types of glazes, particularly earthenware glazes containing lead or tin, are dulled by reduction, and require careful attention, or protection inside saggars, when fired in a fuel-burning kiln. The traditionally highly decorated pottery of Europe, such as maiolica, faience, and slipware, belongs in this category.

Many of the studio potters experimenting with the first electric kilns preferred the reduction glazes and the Oriental aesthetic which were predominant in studio pottery at the time, and found glazes fired in oxidation to be lifeless and unexciting. They used electric kilns for bisque-firing, though some who recognized the advantages of firing with electricity worked on methods of imitating reduction glazes in oxidation. Other potters, through trial and error, began to look for and discover unique and beautiful glazes that require oxidation. Instead of seeing the electric kiln as an easier way to make the same sorts of pottery they always had, they embraced it as a way to expand the boundaries of what is possible with clay and glaze.

The search—though far from over—has been successful. The electric kiln is no longer the poor stepsister to the gas kiln, just a tool to simplify the task of bisque-firing. No longer can potters insist that only reduction firing in gas or wood kilns can result in really beautiful glazes. The years of exploration have resulted in a growing compendium of truly beautiful glaze colors and surfaces that are uniquely possible in a neutral or oxidizing atmosphere. Challenging to achieve in a fuel-burning kiln, especially at stoneware temperatures, oxidation firing has come into its own.

As the electric kiln increased in popularity and ease of use, an industry grew up to provide support to its users. Electric kiln potters needn't formulate their own glazes if they don't want to, or even mix them from recipes. Every year there are more commercial glazes available, for every temperature. In the low-fire or earthenware range, in particular, there is by now a wide variety of glazes, underglazes, pencils, pens, watercolors, and other products, giving potters a virtually unlimited choice of color to be applied in almost every imaginable way. This has been a boon to potters who want to work with earthenware, for the lower the temperature, the more difficult it becomes to compound reliable, easy to apply glazes which are nontoxic to work with and safe for functional ware. Plenty of glazes for higher temperatures are available as well, meaning that electric kiln potters can create fully vitreous stoneware without having to store buckets and bins of raw glaze materials.

The flip side of this ubiquity and ease of use is, of course, that it's easier than ever for people to create objects out of clay without fully understanding the medium. Anyone can buy bagged clay, make things out of it, cover them with premixed commercial glaze and fire them to the temperature the jar indicates in a kiln with a computerized controller. While perfectly good functional pottery and insightful art objects may be created using commercial products and firing them in an electric kiln, the challenge of doing pottery in this way lies in differentiating one's work from everyone else's. Many ceramic artists whose primary interest is in form feel it is a distraction to mix glazes and embrace the commercial products. Other ceramists are primarily concerned with decoration and appreciate these products the same way a painter appreciates commercial oils or watercolors—after all, no one today expects a portrait painter to begin by mixing his pigments!

A third approach, naturally, is to eschew the use of commercial products and mix one's own glazes and other supplies from raw materials. Even this doesn't have to be an arduous process, since by now there are enough glaze recipes available from other potters that electric kiln potters don't really have to formulate their own. The purist, however, feels that creating pottery that is truly unique requires formulating one's own glazes from scratch. This is an exciting and rewarding endeavor, but it takes time away from the production process and requires research into the nature of the materials and their interactions with each other. Any potter can benefit from some exploration in this direction, even if just by trying to understand what others are doing.

The purpose of this book is to help you explore the possibilities and potential of your electric kiln, by providing inspiration for your creative journey. In the following pages you will read about ceramists who rely on their electric kilns to create interesting, unique, exciting, and beautiful work. There are sculptors and potters, and artists who make everything in between. There is work at every temperature range, some of which focuses on bright colors, some on very little color; some focusing on decoration, some on form. For the reader who wants to understand glazes, there is technical information about mixing and formulating glazes. For the reader who wants recipes, there are recipes. For the reader who wants ideas, there are ideas. For the reader who wants stories, there are life stories and technical stories. And, for the reader who simply wants to relax and look at beautiful pictures of art, theare are beautiful pictures. This book exists to help you discover the virtue and versatility of the kiln that brought power to the people.

— Sumi von Dassow

FUNCTIONAL POTTERY—A COLORFUL STORY

The challenge facing all functional potters is to make work with a visually compelling surface that is food safe and easy to clean. For potters working with high-fire reduction glazes, often this challenge is adequately met with a single classic glaze such as a celadon or a temmoku. For potters working with salt glazing or wood firing, the glazing decisions may be as simple as putting a white liner glaze inside and letting the kiln glaze the exterior of the pots. With some kilns the surface color and texture can be highly unpredictable, so an overly elaborate form or extensively applied decoration may not be appropriate.

The challenge for potters firing in an electric kiln is to make the surface interesting, knowing that the kiln won't apply the glaze during the firing or cause the colors to magically alter by means of a changing atmosphere. This challenge is also an opportunity. An electric kiln simply sets the pot's surface but doesn't add to it in any way—thus, the potter is free to embellish it however he wishes. The electric kiln potter may decorate in a variety of ways, confident that applied decoration will not be obscured or altered in unpredictable ways during the firing. At all temperature ranges, electric kiln pottery is characterized by vibrant, rather than subtle, color—whether applied under, in, or over the glaze. Electric kiln potters delight in mingled bright colors, in playful forms accented with translucent glossy glazes, and in decorative brushwork painted over or under white or clear glazes.

Many of the glaze types now used in electric kiln pottery were developed hundreds of years ago for oxidation firing in carefully controlled fuel-burning kilns. Most historic oxidation glaze required lead, and mature in the earthenware range. The introduction of the electric kiln, and the recognition of the dangers of lead glazes, has spurred the development of lead-free earthenware glazes, and boosted research into oxidation stoneware glazes. Such improvements in glaze technology have been a boon to functional potters, as lead-free glazes present fewer hazards to users, and stoneware pottery is stronger and less porous.

Some materials and techniques are as new as the firing technology. Particularly useful to electric kiln potters are commercial ceramic stains, which are carefully compounded mixtures of oxides fired and ground for use in glazes or slips. These stains are highly reliable sources of color most appropriate for use in a neutral firing atmosphere, though creating an interesting surface using stains often requires creative mixing and layering.

The reliability and predictability of the electric kiln means many electric kiln potters feel free to invest a great deal of time making elaborate forms ranging from the playful to the graceful, using a multitude of techniques. Pottery may be painstakingly assembled—from thrown, hand-built and slip-cast elements, and covered with a single, simple glaze—without too much fear that all the work of building it will be lost in the firing.

The following pages showcase several approaches to the challenge and opportunities involved in creating electric kiln pottery. The unifying factor is a commitment to an exciting form and vibrant color. Creating work ranging from the quirky to the elegant, these potters enjoy manipulating both the clay and the glaze to create highly individual work.

PHOTOS: PETER LEE

Teapot, 5 inches (13 centimeters) in height, thrown, altered and assembled mid-range porcelain, multifired in an electric kiln.

Transitions and Transformations

From Reduction to Wood Fire to Cone 6 Electric

by Geoffrey Wheeler

In the late 1970s, I attended a workshop given by Warren MacKenzie, during which he identified himself as a "mud and water" potter (as opposed to a "glaze and fire" potter). This was a differentiation that I remember strongly identifying with. It is the tactile quality of clay that drew me to becoming a potter. Glazing and firing were a necessary part of the process, but they were not what excited me.

While in school, I never glazed more than ten pots at one time. I attended Kansas City Art Institute, and in a program with 70 ceramics majors, you only fired your very best. The concept of using glazes to take the work a step further was not something I was ready to deal with. All I wanted was to keep from ruining my precious few.

Soon after graduation, I set up a pottery and built a 75-cubic-foot reduction kiln. The first time I faced 200 bisqued pots, I was overwhelmed. Over the next few years, I became more comfortable with glazing; however, it remained my biggest challenge. In hindsight, I see that it was a challenge that I avoided, rather than accepted.

My critical thinking ended when the work was leather hard. I simply dipped the pots in clears, celadons, temmokus and Shinos. They were the easy answer, classic tried-and-true solutions, but they did nothing to further the work. I saw the glazes as extraneous, hiding the pots under a skin of glass that clouded my original ideas.

When I headed to the Archie Bray Foundation in Helena, Montana, for a 1994 summer residency, I was positive that access to the wide array of atmospheric kilns would solve my glazing problems. Early American salt-glazed stoneware and wood-fired Japanese tea ceremony ware had been favorites of mine for years. The firing actively continues the visual record of process; instead of a covering skin, the rich, juicy surfaces interact directly with the clay.

I spent three months working with salt-, soda- and wood-fired kilns. During that time of intensive exploration, I produced the strongest work I had ever

Condiment bowls, to 5 inches (13 centimeters) in height, thrown and altered mid-range porcelain, with Cone 6 glazes, multifired.

Bowl, 15 inches (38 centimeters) in diameter, thrown and altered porcelain, with encapsulated-stain glaze, fired to Cone 6, then refired to Cone 04 in an electric kiln.

made, but I soon began to question whether these surfaces were right for my pots. They were becoming too much of a conversation with the past. I realized that the process and history sounded by these surfaces was louder than my own voice as an artist. I decided that, for me, wood- and salt-fired surfaces were just another easy answer, simply another way not to take full responsibility for my finished work.

I had been considering graduate school for a number of years and my experience as a resident at the Bray was the kick in the pants I needed. It was time for me to reassess my work and my life, so I quit my adjunct faculty position, sold off my studio equipment and moved to Minneapolis to work with Curtis Hoard and Mark Pharis at the University of Minnesota.

Grad school is a time for questions. It is not a time for easy answers. I decided to not only "come out" as a gay man, but to come out in my work. I reasoned that coming out was as much about taking responsibility as sexuality. This meant making work that was open and honest. My forms could speak to the relationships of flesh and bones and breath. Color, which is so psychologically charged, could be used to express a sense of joyful exuberance and playful sensuality. I wanted to use glazes to extend the content of the work beyond historical reference, process and utility.

I made hundreds of Cone 10 oxidation glaze tests, developing a palette that alludes to the lush tropical colors that I associate with joy, freedom and transformation. My original plan was to use these glazes in a soda atmosphere to add depth and variation to the surfaces; however, most of my testing was in an electric kiln. I began to wonder if it was possible to get the results I wanted without the hellish experience of loading outdoor kilns in subzero weather. (Yeah, I know, I'm a sissy.) Bright colors and electric kilns sounded like easy answers, but they weren't. The great pots of history, to which I had always looked for answers, were not fired in electric kilns. I had to work out the problems myself.

My previous encounters with high-temperature electric kiln glazes were with

Vase, 12 inches (approximately 30 centimeters) in height, by Geoffrey Wheeler, Menomonie, Wisconsin.

colors that were flat and lifeless, not the lush, rich surfaces that I wanted for my work. The happy accidents that happen in other kinds of firings are not likely to happen in electric kilns. Electricity doesn't carry the magic kiss of the flame. But I found that by using similar colorants in both matt and glossy glazes, and by spraying and overlapping the two, I could build surfaces with the kind of visual depth and variation I was looking for.

In addition, I discovered that while some of my glazes had the typical flat quality of electric firing at Cone 10, when refired to Cone 04 they developed unexpected color and surface variations. Turquoise and purples emerged in areas that had previously been green (from copper and nickel). Some of the commercial stains that start to fade out at Cone 10 reasserted themselves during the second firing.

In the past year, I have reformulated my Cone 10 base glazes to mature at Cone 6. I am getting almost identical results, while reducing the wear and tear on the kiln. I leave some of the exterior areas of the pots unglazed during the initial Cone 6 firing and apply Cone 04 glazes that differ in surface quality and color intensity for the second firing. Multiple firings, using a variety of types of glazes, can come together to build a palette of extraordinary richness and depth. Glazes with chrome sometimes flash onto other surfaces, giving a taste of serendipity in an otherwise controlled atmosphere.

What most interests me about working within the vessel tradition is how we perceive these voluminous forms in relation to ourselves. Whether they are viewed in terms of function and physical interaction or in terms of symbol and metaphor, we identify with them as extensions and representations of ourselves. While I used to think that glazes distracted from the essence of the pots, hiding the primal forms underneath, I now feel able to use them to continue the ideas, reinforce the forms, further the content and perhaps transform them from reality to symbol and back. ▲

Recipes

The following glazes are used on pots made from a Grolleg porcelain produced by Standard Ceramics of Pittsburgh:

Glassy Alkaline Glaze
(Cone 6)

Gerstley Borate	1.19%
Lithium Carbonate	3.38
Strontium Carbonate	9.30
Frit 3110 (Ferro)	21.12
Kona F-4 Feldspar	46.34
Flint	18.67
	100.00%
Add: Bentonite	3.00%

This glaze begins to flux at Cone 04, but I have used it as high as Cone 10. It runs easily at the higher temperatures and tends to craze badly when thick. For a water-blue variation, add 1.00% copper carbonate; for yellow, add 5.00% zirconium yellow stain or 5.00% vanadium yellow stain.

Rob's/G.A. Blend Glaze
(Cone 6)

Gerstley Borate	2.9%
Lithium Carbonate	1.7
Strontium Carbonate	9.3
Whiting	8.4
Cornwall Stone	34.6
Frit 3110 (Ferro)	10.6
Kona F-4 Feldspar	23.2
Flint	9.3
	100.0%
Add: Bentonite	3.0%

A waxy, semitransparent glaze. For a deep blue variation, add 2.0% cobalt carbonate and 4.0% manganese carbonate; for warm pink, add 5.0% coral stain and 3.0% rutile; and for apple green, add 5.0% green stain. Manganese should be handled with care and fired in a well-ventilated kiln; avoid inhaling dust or fumes.

Matt "B" Glaze
(Cone 6)

Lithium Carbonate	2.7%
Strontium Carbonate	26.5
Nepheline Syenite	57.5
Kentucky Ball Clay (OM 4)	6.2
Flint	7.1
	100.0%
Add: Bentonite	3.0%

The matt quality of this glaze is easily affected by colorants, so variations have different surface qualities as well as colors. For a dark green variation, add 3.0% copper carbonate and 2.0% nickel oxide; for orange, add 5.0% encapsulated orange stain and 5.0% zirconium yellow stain; for maroon, add 5.0% coral stain and 3.0% manganese carbonate. I use Degussa encapsulated stains; to be considered food safe, these must be prepared/used in accordance with the manufacturer's instructions.

The Evolution of an Idea

RAY BUB'S REASSEMBLED RING TEAPOTS

by Paul Park

The teapot has fascinated ceramics artists in both the East and the West for the past 600 years. It is the queen of pottery shapes, a formal puzzle with limitless solutions. As a result, the challenge of making something new, a unique and compelling functional teapot, is a demanding one.

Ray Bub has been making pottery and teaching students at his studio in Pownal, Vermont, for 34 years. He had been intrigued with the teapot format for some time before he took his Southern Vermont College class to the Bennington Museum to see the pottery collection in the spring of 1995. After that study trip, one student, Dylan Lawson, mentioned that he would like to make a ring vase similar to the 18th-century ring flask (made to fit around a man's forearm) in the museum's American folk pottery display. Bub showed him how to throw a hollow ring, then attached an oval base and a bottle neck. Afterward, the demo remained in the studio.

Soon, Bub started thinking about adding a spout, handle and lid to this traditional ring vase shape. The resulting teapot had an elegant and pleasing form, but was almost mute in its completeness. Bub's first instinct was to give it a voice by adding some of the hand-sculpted animal figures he'd been putting on boxes since 1992. So he added a spotted jaguar to the lid, and perched another on the inside surface of the upright ring. He then made several upright ring teapots with different animals.

It was in the fall of 1995 that he made his first reassembled ring teapot. Again, there was nothing planned about the process: when the hollow ring was at the leather-hard stage, he had cut it apart with a bow saw, planning to reverse a couple of sections to create a zigzag profile to the upright ring. But the open ends of the cut-apart sections were unmatching trapezoid shapes that would not reassemble into a symmetrical closed form. Unhappy that he'd ruined the ring and wasted the time he'd spent on it, he decided to try to salvage his investment, and began rearranging the arc sections in different ways. Immediately, he was intrigued by the visual possibilities. He joined the arc sections together end to end and out of order, then put flat slabs on the two open ends. Then he added an oval base, a spout, a handle, a neck opening and a lid, embellished with a keel-billed toucan.

PHOTOS: JON BARBER, SUSAN NYKIEL

"Keel-Billed Toucan Reassembled Hollow-Ring Teapot," 15 inches (38 centimeters) in height, wheel-thrown, cut and assembled stoneware, fired to Cone 5 in oxidation.

From 1995 to 1998, he made numerous reassembled-ring teapots, all decorated with animal shapes—Madagascar chameleons, African elephants, African giraffes, king penguins, highland gorillas, Pacific puffins, ring-tailed lemurs, North American mountain goats, etc. He made some sales, but after paying commissions, there was not a great return on the time invested, so he continued to earn most of his income by teaching classes in the studio, and making and selling "conventional" functional ware.

During this period, he also submitted slides of his animal figure teapots to several prestigious national and interna-

"Pink-Green Oval Cross-Section Reassembled Hollow-Ring Teapot," 14 inches (36 centimeters) in height, thrown and assembled stoneware, fired in oxidation to Cone 5.

First, a ring is thrown by joining two walls at the top, trapping air inside.

When the ring is leather hard, it is inverted and trimmed.

Sections are then cut at various angles using a bow saw.

The section is closed with slabs, traced and cut to fit each end.

When reassembly is complete, clay spacers and supports are added for stability during drying, then the form is positioned on a thrown oval base.

A handle is pulled from a lug attached to a ring section, then a thrown spout is shaped and attached.

The last step is to make a finial for the lid; several are made and the one that most successfully enhances the design is attached.

tional craft fairs and exhibitions, but was not accepted to the ones he most wanted to participate in. As time went by, he grew more dissatisfied. Did the animal figures he had been attaching to the upright-ring and reassembled-ring teapots somehow cause jurors not to choose his work? Put another way, did the embellishment distract from the integrity of the design? When discussing this work with customers, he found them referring not to the teapot composition itself, but to the animals, with such comments as, "I love the chameleons," or "Elephants are my favorite animal."

An artist frequently encodes ideas in some form to tell a story, but it was clear Bub's story wasn't coming through. He decided to put his trust in the intrinsic, undecorated eloquence of the abstract forms, and in the spring of 1998 made his first reassembled ring teapot without animal figures. It was his "Pink Pentagon Cross-Section Teapot," and it was one of 112 works selected for the 1999 "Ceramics Monthly International Competition." The jury had chosen it from among 1549 slide entries from 45 countries. "Wow!" he thought. "No more animal figures for me."

Soon he began experimenting with round, square, pentagonal, distorted, oval, trapezoidal and star shapes, cutting them into various-length arc sections and reassembling them into balanced compositions, positioning the assemblage on a thrown base, cutting out the lid, then adding spout, handle and finial. Since the summer of 1998 to September 2001, he has made 24 abstract reassembled hollow-ring teapots and two upright-ring teapots, all without animal figures. He has always used all the sections from one complete ring in each teapot's assemblage, and has added the same elephant-trunk spout and pulled handle, although their placement varies depending on the composition.

"Orange Five-Pointed Star Cross-Section Reassembled Hollow-Ring Teapot," 19 inches (48 centimeters) in length, wheel-thrown, sectioned and reassembled stoneware, glazed and fired to Cone 5 in oxidation.

"Silverleaf Bonsai Tree Reassembled Hollow-Ring Teapot," 11 inches (28 centimeters) in height, fired to Cone 5 in oxidation.

Twelve of these abstract reassembled-ring teapots have sold (through various venues) for $1200 each, less an average 33% commission. One of the most interesting sales opportunities has been the Eziba website, which has sold five of his teapots since the spring of 2001 to collectors in Dallas, Texas; Aspen, Colorado; San Antonio, Texas; Williamstown, Massachusetts; and Toronto, Ontario, Canada.

Meanwhile, Bub has continued sending out slides to competitions, with similarly positive results: His "Celery Green Reassembled Hollow-Ring Functional Teapot" was one of 151 pieces chosen by three international jurors and three Taiwanese jurors for the "Sixth Taiwan Golden Ceramic Awards Competition" in the spring of 2000. It was awarded the "Special Judge's Prize" from Taiwanese juror Bob Chen; and the Taipei County Yingko Ceramics Museum, the host and one of the principal sponsors of the exhibition, purchased this teapot for its permanent collection.

During the winter of 2000, he also sent slides to the "First World Ceramic Biennale 2001 Korea International Competition." His "Lemon Yellow Triangular Cross-Section Reassembled Hollow-Ring Teapot" was one of 305 works by artists from 69 countries selected from 4206 entries. The exhibition was presented at the Ichon Ceramic Center as part of the "World Ceramic Exposition 2001 Korea," a massive, $100 million ceramics world's fair, which drew an estimated five million visitors. This multi-event celebration of historical and contemporary ceramic art from around the world was the largest event of its kind to date.

This recognition is, of course, gratifying to someone who for more than three decades has been making functional pottery and gives them tension and fluidity. Often the forms shift in firing, in ways that cannot be anticipated. Typically,

"Grasshopper Leaping Reassembled Hollow-Ring Teapot," 16 inches (41 centimeters) in height, by Ray Bub, Pownal, Vermont; www.raybub.com.

Bub lives with the leather-hard reassembled ring composition for a while before he chooses the location of the base, the spout, the finial—all the details that will solidify the piece in the viewer's eye.

In the case of his "Orange Five-Pointed Star Cross-Section Reassembled Hollow-Ring Teapot," he "had originally imagined this was going to be a tall piece, but when I was unwrapping it, I thought immediately of the Taiwanese ceramist Ah Leon and a teapot he had made in the shape of a long tree branch, which is in the collection of the Metropolitan Museum of Art in New York City. Thinking of this Yixing-inspired teapot, I laid my reassembled ring composition on its side, then defined it by the location of the handle and spout. It needed no base, as it turned out. But to maintain functionality, I had to turn the spout upward."

Since Bub began working with this form, he has seen an earthenware ring vase with an oval base, made in Apulia in central Italy in 340 B.C. He has seen ring vase examples or pictures from the Japanese Kofun dynasty, the Chinese Tang, Song and Ming dynasties, ninth-century Moorish Spain, and twelfth-century Persia, as well as the colonial pilgrim flask in the Bennington Museum. Though geographically and chronologically widespread, the form is still a rarity, which makes it possible to deconstruct it and reinvent it for a long time and in myriad ways, without approaching anyone else's work. Bub believes there is boundless territory for him to explore with just this one vehicle. As for the next idea, he's preparing for another accident. ▲

Glaze Recipes

From Ray Bub

Glossy Clear Base Glaze
(Cone 5)

Whiting	4.00%
Frit 3195 (Ferro)	40.50
G-200 Feldspar	16.50
Bentonite	2.50
Edgar Plastic Kaolin	6.50
Flint (400 mesh)	30.00
	100.00%

Color variations are possible with the following additions:

Opaque White:	
Zircopax	15.00%
Translucent Pink:	
Whiting	10.00%
Mason Stain 6002	6.00%
Translucent Pale Purple:	
Mason Stain 6332	4.00%
Translucent Dark Red-Purple:	
Whiting	10.00%
Mason Stain 6003	5.00%
Mason Stain 6338	0.50%
Opaque Pale Blue:	
Zircopax	15.00%
Mason Stain 6364	3.00%
Translucent Pale Blue:	
Cobalt Carbonate	0.75%
Iron Oxide	0.75%
Translucent Dark Blue:	
Cobalt Carbonate	2.00%
Iron Oxide	1.00%
Translucent Teal Blue:	
Mason Stain 6371	0.75%
Translucent Yellow:	
Mason Stain 6450	6.00%
Translucent Orange:	
Mason Stain 6450	5.00%
Light Tone Rutile	10.00%
Translucent Pale Green:	
Copper Carbonate	2.00%
Translucent Dark Green:	
Copper Carbonate	6.00%
Translucent Pale Brown:	
Manganese Dioxide	4.00%
Translucent Dark Brown:	
Manganese Dioxide	8.00%
Mason Stain 6600	0.50%
Black:	
Mason Stain 6600	8.00%

A base matt glaze was adapted from a recipe given by Father Anthony Bellasorte in the February 1994 CM.

Bronze Tan Matt
(Cone 5)

Whiting	21.51%
Zinc Oxide	9.14
G-200 Feldspar	50.54
Edgar Plastic Kaolin	12.90
Flint (400 Mesh)	5.91
	100.00%
Add: Bentonite	2.15%
Light Rutile	5.38%

For Bronze Green Matt, add 0.5% cobalt oxide and 3% copper carbonate; for Bronze Blue Matt, add 1% cobalt oxide.

I use commercial stains to lessen the pitting and bubbling common to low-fire glazes colored with oxides and carbonates. The use of stains also widens the choice of colors. I opacify most of the Cone 06 glazes for true colors on the dark red earthenware clay we use, as its fired strength is superior to white or buff earthenware bodies. The red earthenware is also more plastic and forgiving, an important consideration for beginning potters.

Glossy Clear Base Glaze
(Cone 06)

Frit 3124 (Ferro)	85.0%
Edgar Plastic Kaolin	15.0
	100.0%

Color variations are possible with the following additions:

Opaque White:	
Zircopax	16.0%
Opaque Pink:	
Zircopax	11.0%
Mason Stain 6001	6.0%
Opaque Purple:	
Zircopax	11.0%
Mason Stain 6331	5.0%
Opaque Red-Crimson:	
Zircopax	11.0%
Mason Stain 6003	5.0%
Opaque Yellow:	
Zircopax	11.0%
Mason Stain #6450	5.0%
Translucent Orange-Mustard:	
Mason Stain 6450	5.0%
Light Tone Rutile	10.0%
Opaque Light Green:	
Zircopax	11.0%
Mason Stain 6271	1.0%
Translucent Dark Green:	
Mason Stain 6271	2.0%
Black Mason Stain #6600	0.5%
Opaque Pale Blue:	
Zircopax	11.0%
Mason Stain 6364	3.0%
Opaque Medium Blue:	
Zircopax	11.0%
Mason Stain 6313	1.0%
Translucent Dark Blue:	
Mason Stain 6386	1.0%
Translucent Brown:	
Iron Oxide	2.0%
Black:	
Mason Stain 6600	8.0%

Stoneware vase with cane handle, 10 inches tall, and stoneware coffee mug, 4 inches tall, by Susan Nykiel; stoneware vase, 9 inches tall, by Ray Bub.

Ray Bub Explains His Glazes:

All the glazes depicted in this article are cone 5 oxidation glazes made from the recipes on the glaze page, with the exception of the last photo, Grasshopper Leaping Reassembled Hollow Ring Teapot. The glaze on this teapot is a cone 10 reduction scrap glaze with no fixed recipe.

The Pink Pentagonal Teapot had "Translucent Pale Green" sponge-printed on the edges, then "Translucent Pink" poured over the whole teapot. There was "Translucent Yellow," "Translucent Dark Blue," and "Translucent Pale Green" painted into the fissure grooves and on the edges of the lid finial before a second coat of "Translucent Pink" was poured over the lid.

The Pink-Green Oval Teapot, had "Translucent Dark Green" painted into the grooves in the oval ring sections and on the sides of the zigzag lid finial, then a watery thinned-down recipe of "Translucent Pale Green" poured over the whole teapot and (separately) over the lid. A second coat of "Translucent Pink" was then poured over the whole teapot and over the lid.

For some chemical reason that a ceramic glass scientist (not me!) may be able to explain, the copper carbonate in the "Translucent Green" first layer burns through and half-neutralizes the chrome-tin pink color of the "Translucent Pink" second glaze layer. The results are variable pink/purple/green depending on where the green and pink glazes are thicker and thinner.

The glaze on the Keel-Billed Toucan Teapot is the cone 5 "Bronze Green Matt." The glaze on the Orange Five-Pointed Star Teapot is "Translucent Orange" from the glaze page poured over the teapot and lid. I have been getting some bubbling and pitting at cone 5 with this recipe, which has 15% colorant additions, and I have recently substituted an expensive but no-pitting addition of 5% Orange Mason Stain 6024 to the cone 5 Glossy Clear Base Glaze with good results.

The clay body used for the Pink Pentagonal Teapot, the Pink-Green Oval Teapot, the Keel-Billed Toucan Teapot, and the Orange Five-Pointed Star Teapot, is Sheffield Pottery's recipe "T-3 Screened cone 10" slightly underfired to cone 5 in our electric kiln. The Grasshopper Leaping Teapot is made with this same T-3 Screened clay body, but fired to cone 10 in our propane-fired reduction kiln.

The Silverleaf Bonsai Tree Teapot has "Opaque White" poured on the whole piece as a first layer, then "Translucent Pale Purple" poured on as a second layer. The clay body for the Silverleaf Bonsai Tree Teapot is Sheffield Pottery's "T360M cone 5" recipe fired to cone 5 in our electric kiln

Shuji Ikeda

by James Irwin

As we stepped into the entryway of Shuji Ikeda's house in Berkeley, California, my wife and I were drawn to a large, handsome arrangement of irises with tiny white blossoms, a type of iris known in Japanese as *shyaga*. The assembly sprouted from a simple but dramatically flared vessel thrown from a rich, dark clay body.

The evidence of Shuji's dual passion for clay and flowers—and the context of his Japanese heritage—was visible everywhere. In one corner was a collection of vessels designed explicitly for flower arranging, or *kado*. Some were thrown; some were handbuilt. The variety of sizes and shapes seemed endless.

We were there on an East Bay Open Studio day, which meant we had the opportunity to observe the reactions of many visitors. A group of elderly Japanese women hovered over the kado pieces. From the tone of the animated but indecipherable conversation, it was clear that this is serious business. Indeed, Shuji's kado pieces are highly sought after in the flower-arranging community both in the San Francisco Bay Area and in Japan. He views the line of kado pieces more as a vehicle for experimentation and play, however. His favorite form is the handbuilt basket.

In front of the window, bathed in full daylight, was a basket display. These forms blend a solid architectural stateliness with an intricate texture of coils and strips of clay that have been wrapped, braided and woven, or fashioned into delicate twigs and leaves.

Some of the baskets are glazed with a blue-green matt glaze, which Shuji calls *Sei Shya* (Blue Rust). Others are sprayed with iron or manganese oxides. Many have no surface treatment, but instead show off the dark, smoke-colored clay body from which they are constructed. In some, the dark body has been combined with a red clay by partial wedging, a traditional Japanese technique known as *nerikomi*. The degree to which the clays are wedged together results in varying effects when the clay is cut into

"Mum Leaves Basket," 17 inches in height, slab and coil built, with Blue Rust Glaze, fired to Cone 5.

PHOTOS: SHUJI IKEDA, RICHARD SARGENT

"Woven Ceramic Basket," 17 inches in height, constructed from red and black clays, clear glazed.

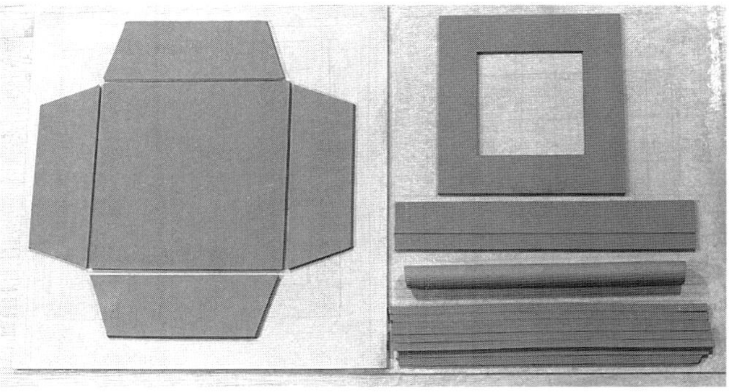

All parts are measured and cut from slabs and extrusions.

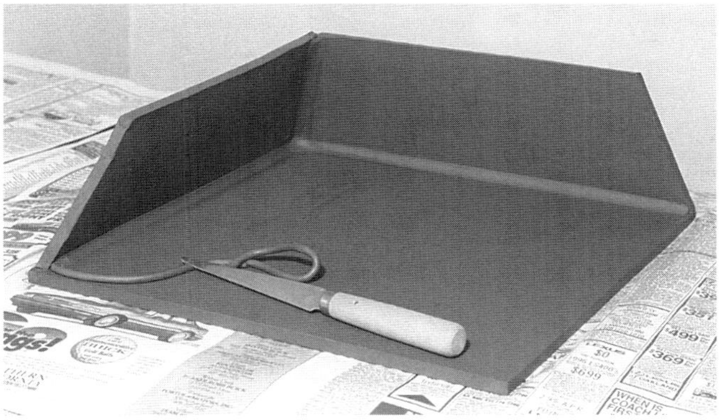

Joins are reinforced with extruded coils.

Once the walls are assembled, the top slab is attached.

strips or rolled into coils, then braided or wrapped.

I have been watching the evolution of Shuji's baskets for several years now. Earlier versions were direct interpretations of Japanese flower-arranging baskets known as *hanakago,* which are constructed from twigs, reeds or split bamboo. His newer work shows a more personal touch. "I am trying to create a kind of metaphor by mixing two ways of mimicking nature," he explains.

One way is to use the natural character of the clay—how it rolls, twists, breaks and bends. The other is the introduction of trompe l'oeil natural objects—twigs and leaves. Shuji calls this work *tsuchi kago,* literally "clay basket."

The story of how these pieces came to be is the story of how a Japanese immigrant became a potter in America. The surprise twist is that Shuji's pottery teachers were not Japanese, but American. He arrived in the U.S. at the age of 23 (and likes to point out that he has now lived here more than 23 years).

He completed film studies at San Francisco State University, but because jobs in that field were scarce in northern California, he went to work selling Asian antiques for Sloan Miyasato at the Design Center in San Francisco. He was hired because he had translation skills, and was knowledgeable about Japanese pottery (which he had collected in Japan) and Asian antiques in general.

Sometime in the early '80s, he wandered into Pottery 7, the cooperative ceramics studio located in the Inner Sunset district of San Francisco, and

Next, the legs are attached and reinforcement strips are added to the corners and bottom edges.

Notches are cut from the top for the extruded handle, and extruded coils are wrapped around the corners.

The coils are attached one by one.

The handle is wrapped with long coils.

signed up for lessons. His American teachers were somewhat amused to find themselves in the role of mentors to someone from the culture where ceramics is revered more than anywhere else in the world. Indeed, the work he began marketing through craft fairs five years later echoed classical Western shapes, and its only suggestion of Japanese influence was the raku firing.

Shuji traces his renewed interest in his native heritage to two events: moving into his own studio, and a decision to study kado. He explains that his kado teacher arrived in America in the early '50s and is thus more connected with older Japanese culture.

"My approach to learning kado was typically American," he says. "I told her to 'teach me everything in three sessions.' She laughed, gave me a bulb and told me to go home and plant it. Several months later, I cut the flower grown from the bulb and took it to class. She asked, 'Did you see how the flower broke the ground?' No. 'Did you see what the weather was like when it broke the ground?' No. 'Then how can you know how to arrange it?'"

Seven years later, he is still studying with the same teacher, and credits this experience with kindling a passion for "mimicking nature" in clay. He feels especially fortunate to have become immersed in two very different disciplines of artistic endeavor. "In Japan, if I had wanted to be a potter, I would have had to spend two or three years sweeping the floor. Sacrifice and hardship are valued highly. Discipline is traditional and external. There are rules. Manners and morality are important. In America, individual rights are important—you don't

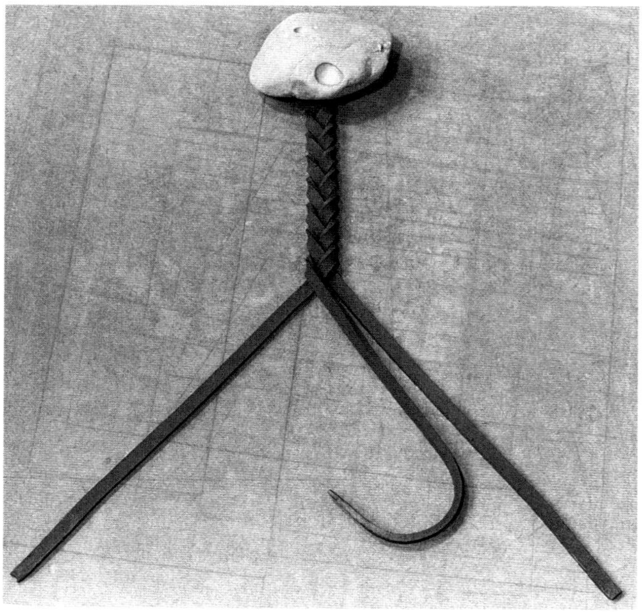

Flat coils are carefully braided for side insets.

The braids are measured and cut to fit precisely.

"Woven Ceramic Basket," 12 inches in height, clear-glazed black clay.

do anything you don't want to. Discipline is more internal, driven by the passion for what interests you. There is no 'way.' There are no rules. I struggled with that in pottery, always asking, 'What is the right way?' Until one teacher said, 'I don't care what you do or how you do it. Just make pieces that are beautiful to you.'"

Pointing to the shyaga arrangement, Shuji explained the functions of the various parts of the composition: "heaven," "earth" and "human" combined to express the theme that all creatures are supported by heaven and earth, which allows them to grow. He then showed me a book of "rules" just for irises; different types of irises have different rules, all the accumulation of a thousand-year-old tradition of flower arranging.

Later, he pulled out a photograph of a red flower seemingly tossed in a depression in white snow. "This throws out the rules," he said. "When you understand the rules, you can throw them away, but not until then."

In a way, both disciplines—Japanese and American—are different approaches to this same end, he remarked. "In my search for identity, I like to watch the interplay of both."

Before departing, I paused to take in another arrangement, this one in a clay basket. In contrast to the traditional hanakago, Shuji's baskets evoke a more architectural, monumental stateliness. In this way they stand alone. However, this hasn't stopped him from creating flower arrangements in them.

As a former ceramics artist, I had developed a prejudice against seeing anything put in my pots, or anyone else's. Yet as my eye wandered between the pink blossoms and the basket that held them, I began to wonder if the arrangement might not be the vessel for the ceramic basket. And if perhaps this wasn't how these traditions of discipline keep each other alive. ▲

Recipes

Blue Rust Glaze
(Cone 4–5)

Barium Carbonate	32.55%
Dolomite	23.26
Custer Feldspar	37.21
Edgar Plastic Kaolin	6.98
	100.00%
Add: Copper Carbonate	2.33%

Clear 3B Glaze
(Cone 5)

Gerstley Borate	47.5%
Edgar Plastic Kaolin	24.2
Flint	28.3
	100.0%

Shuji Ikeda in his Berkeley studio.

A Dialogue with Lisa Orr

by Rafael Molina-Rodriguez

Pitcher, 8 inches in height, earthenware, wheel thrown and altered, with sprigging, rouletting, slip trailing, terra sigillata and multiple glazes, fired to Cone 03 in oxidation.

Texas potter Lisa Orr produces colorfully glazed wheel-thrown and altered earthenware teapots, pitchers and bowls, as well as pressed and altered platters. This interview was conducted at her studio adjacent to her home in Austin.

Interviewer: How did you get started in clay?

Orr: I discovered ceramics in college. I went to the University of Texas at Austin and received a bachelor of fine arts. I then worked for local potters Ishmael Soto and Dee Buck to build up my experience—I would work two or three days for them and two or three days for myself. I had a little store in San Antonio for about three years, and I did craft fairs. At the same time, I took a whole bunch of workshops, and decided I needed more education.

I didn't think I wanted to go to graduate school. I thought it would be better to study with someone in particular, so I went to the University of Colorado and studied with Betty Woodman as a special student in 1989. Next, I went to the University of Minnesota where I spent nine months as a special student, studying with Mark Pharis.

My work was self-conscious, but for whatever reason, I needed to do it. We need to go through these self-conscious periods, rather than avoiding them, in order to get to something else. So I was probably making some very self-conscious terrible pottery, but it was what I needed to make at that time. It was sort of a pseudo folk-craft mingei-sota style or, at least, my interpretation of it.

I then went to graduate school at Alfred. It was a very nurturing and encouraging environment. All the doors that I didn't even know were there were flung open: the library, museum and the clay room. I tried everything that I could while I was there.

Interviewer: What effect did the time you spent with Betty Woodman have on your work?

Orr: It was very freeing. I was very interested in historical ceramics, but I had felt guilty borrowing from them. I felt like I should somehow be more original, or this was not from my time period and I shouldn't use it as a source.

One question that she always asked the students was, "Why not?" And so I started asking myself, "Yeah, why not?" It freed me to enjoy making those things and then move beyond them in my own way.

In my attempt to make good copies of ancient pots, I always wound up making something else. I also invented a lot of interesting little tools to attempt to make them.

Interviewer: What are some of the other influences on your work?

Orr: All kinds of folk pottery, including French, Oriental and Mexican, as well as restaurant ware. Mexican pottery, however, is my number-one influence. Some things, like the "Trees of Life," are done with such freedom and in a casual unself-conscious manner. There isn't anything that's less self-conscious. Mexican pottery also has the added ingredient of humor.

Interviewer: Prior to establishing your studio in Austin, you traveled extensively. What effect did your sojourns in Eastern Europe have on your work?

Orr: I first went there in 1991 because of my interest in international folk dancing. Every five years there is an international folk-dancing extravaganza in Bulgaria. This trip was between my first and second year of graduate school. While I was there, I also toured Macedonia and Albania. I looked for potters, and made contacts. When I returned to Alfred for my last year of grad school, everyone encouraged me to apply for a Fulbright. I did and a year later I went. I received a $13,000 stipend for travel, food and lodging.

I was based in Bulgaria from September to May. I was interested in studying the folk pottery of the region, and was loosely associated with a university as their student, but the faculty had no time or interest in my project.

I first went to the village of Etur, where there was a woman who made very nice folk pottery. She was out of the postwar generation; before the war, women didn't make pottery. After the communists took over, both men and women potters were trained in schools, rather than in the guild system.

I also traveled to Macedonia, which was far more interesting to me. In 1991, as a part of Yugoslavia, it was enmeshed in a civil war, and Fulbright would not send people there. Still, I traveled there as often as I could because the communist government hadn't changed (for better or worse) the old ways of making pottery and doing business.

Bulgaria was a mixture of old and new ways. In Bulgaria, potters are required to use electric kilns, but the energy supply was unreliable. There were unpredictable blackouts when energy was in short supply.

Tureen with tray, 7 inches in height, wheel-thrown and altered earthenware, with sprigging, rouletting and slip trailing, dipped in terra sigillata and brushed with multiple glazes, fired to Cone 03 in oxidation.

Interviewer: What was of particular interest to you about the Balkan potters?

Orr: Several things. I was really interested in the low-fire wood kilns that they used in Macedonia. The Roman type of kilns, which are basically a campfire with a floor above it, with holes poked in the floor and the top part of the kiln enclosed. These are large and they fire very quickly. They use thin light saggars with broken pieces for shelves and tumblestacking.

Interviewer: Do they tumblestack the glazed pots?

Orr: Yes. They're glazed. It doesn't seem so amazing to some people that they fire glazed greenware, but it is to me. And when they're in a hurry, they raw glaze and stack an entire kiln full. They can fire to temperature with scavenged dead wood in about three hours.

At 8:00 the next morning, the women were out there unloading a firing that ended the previous night at 10:00. It seemed too fast, yet the glazes were shiny with hardly any flaws.

Interviewer: What did you learn about the potters?

Orr: I wanted to know why somebody would do that job. In a village of potters, are they all marvelous and unself-conscious? Is it out of love for what they're doing or a great passion for form, or is it simply a need to have something to barter? Have they been made the same way since time immemorial?

Interviewer: And what did you discover?

Orr: I found out that a lot would do anything other than pot. For them, it's a very filthy, dirty life. In the village of Macedonia, they still made traditional items to sell to the locals and tourists; however, because of the war, their market was decreasing.

Interviewer: Let's talk about your work. You utilize a number of forming and decorating techniques. Will you elaborate on some of the materials and processes you employ?

Orr: Some things are thrown, then altered. Although I really like some of the effects that I can get from throwing, I'm looking for other effects. I like the combination of mold-made parts and thrown parts together. I'm trying many different ways of doing it. For example, I hadn't made pitchers in about a year because I'd lost my passion for making them the way I used to. Recently, I made a mold that I throw the body in upside down, then add a spout.

Interviewer: In addition to throwing and altering, you do some unique hydraulic pressing.

Orr: It's different than anyone else I know. I'm using the press to get a beautiful edge. I put too little clay into a too large mold. I place it in certain shapes so that it fills the mold a certain way. Also, before I completely release the clay out of the mold, I can model it with roulettes. I cull the ones that don't fit my aesthetic.

Interviewer: Will you elaborate on your decorating techniques?

Orr: Rouletting is one. My stamping and embossing tools are made of fired clay; I use some soaking wet and some dry. I also use slip trailing and sprigs.

I dip the work in a terra sigillata made from ball-milled Alfred shale. Part of it goes away underneath the glaze, but I love the depth I get with it. I have different glazes that I overlap in different ways. It's amazing how long it takes me to learn how to use combinations from these six glazes well. They're thick and viscous like Elmer's Glue and I apply them with brushes.

Interviewer: How do you fire?

Orr: In an electric kiln. A lot of them are fired on edge. Dennis Olson, the ceramics instructor at Amarillo College, made some stainless-steel plate racks for me. I can fire lots of plates on one shelf this way.

Interviewer: What temperature?

Orr: Bisque to Cone 01 and glaze to Cone 03 in oxidation.

Interviewer: Will you discuss some of the formal aspects—i.e., color, contrast, pattern, texture—of your work? Does low-temperature firing make color easier for you?

Orr: Yeah, for me it does. Although I've seen people do it successfully at higher temperatures, I cannot. And I don't want to. I don't want to use a gas kiln because right now I don't have a passion for high-temperature work. I work hard, glaze playfully and stack well. I don't want to watch the kiln firing. I am content with the slightly more fragile low-fire things.

Interviewer: When you're designing pieces and applying the sigillata over the clay body, do you have in mind areas that you want contrasts of lights and darks?

Orr: Yes. Right now, I want to have less light and more dark. I don't use too much white, except on some pieces I want completely white so you can see the form better.

Interviewer: Texture is obviously an important element to you as well.

Orr: It's my favorite aspect, though I think my best work doesn't need it. Sometimes, on the best ones, I'll hold back and lightly touch them with decoration, perhaps just one sprig. I'll glaze it one single color with maybe some splashes of another color.

Interviewer: Asymmetrical balance is another element in your work.

Orr: It all started with Tony Hepburn asking me the question in graduate

Platter, 10 inches in length, pressed and altered earthenware with sprigging, rouletting and slip trailing, dipped in terra sigillata, brushed with multiple glazes, fired to Cone 03 in oxidation.

Bowls, 8 inches in diameter, wheel-thrown and altered earthenware with sprigs, rouletting, terra sigillata, slip trailing and multiple glazes, fired to Cone 03 in oxidation, by Lisa Orr, Austin, Texas.

school. "Why symmetrical?" I couldn't answer it then, and now I can. I've certainly thought about it a lot. The questions I ask myself now are: "Why not asymmetrical? Why not symmetrical? Why not just whatever is right for the moment?"

Interviewer: It's very organic and curvilinear.

Orr: Yes. It just keeps my eye on the piece and engaged. Hopefully, the viewer's eye as well. And there is something about the undulation, the sensuous flow of a wavy edge that I love more than maybe a straight, flat round one.

Interviewer: Is that part of your idea with the hydraulic press, to create that undulating rim or lip?

Orr: Yes. And they're all different when they come out. I can change them, too; they're still soft and malleable.

Interviewer: Your recent work has gone through an evolution. You have moved away from the functional to more sculptural forms. What are your thoughts about utilitarian pottery versus decorative vessels?

Orr: Back in the 1980s, my work was more functional. It was easier to sell because people understood it. They had seen it before. I made a lot of stackable bowls, plates and things like that. Now, I'm following what I really want to make in my heart, and not necessarily what the public easily recognizes and understands. They might find an unusual salad bowl over here and it may be the only one like it that I'll ever make.

Interviewer: So the work you're doing now is more one-of-a-kind?

Orr: They come in series. Sometimes I'll just decide I'm going to make a little batch of compotes a certain way. I'll make a new mold then. All of them will be very different in form and mood.

Whenever I look at a mold and have that slight feeling of dread, I just break it so I don't have to see it anymore. There's plaster in the driveway right now. This opens the door for new forms.

Interviewer: What plans do you have for your work in the immediate future?

Orr: I need to work on a more successful pitcher in a mold, and invent a new teapot and casserole. Right now, I'm taking little baby steps, and I'm getting awkward results, because it's the first time to do something very new.

Marketing wise, I really want to sell one-of-a-kind pieces online. I've been working with someone on my website (www.lisaorr.com). When it's completed, it will be ideal for me, because my entire inventory will be on view.

The website should make it easier for my wholesale accounts to order, since these pieces can be so different from one another. The pieces shown on the website will be for the galleries to see first, and for me to retail whatever they do not select.

Marketing online should be great, because I will have the freedom to make something new each time I go into the studio, rather than take orders. ▲

Glaze Recipes

Clear Base Glaze 1
(Cone 03)

Gerstley Borate	9.00%
Soda Ash	16.00
Frit 3110 (Ferro)	66.00
Edgar Plastic Kaolin	5.00
Flint	4.00
	100.00%

Color variations are possible with additions, such as the following:

Honey Yellow:	
Spanish Red Iron Oxide	11.00 %
Turquoise:	
Copper Carbonate	6.00 %
Soft Blue:	
Cobalt Carbonate	0.50 %
Purple:	
Manganese Silicate	5.00 %

(Note: when using this colorant, delete the silica in the base.)

Green:	
Chrome Oxide	0.25 %
Copper Carbonate	6.00 %
Red Iron Oxide	11.00 %

Better color variations are achieved with Clear Base Glaze 1, but Clear Base Glaze 2 is more stable.

Clear Base Glaze 2
(Cone 03)

Gerstley Borate	12.00%
Frit 3110 (Ferro)	73.00
Edgar Plastic Kaolin	15.00
	100.00%

Color variations are mixed with these additions:

Honey Yellow:	
Spanish Red Iron Oxide	11.00%
Turquoise:	
Copper Carbonate	6.00%
Soft Blue:	
Cobalt Carbonate	0.50%
Purple:	
Manganese Silicate	5.00%
Green:	
Chrome Oxide	0.25%
Copper Carbonate	6.00%
Red Iron Oxide	11.00%

Jared Jaffe's Functional Fantasies

by Lisa Hurwitz

"Pumpkin Teapot," 11 inches in height, slip-cast and handbuilt whiteware, with Cone 06–04 glazes.

Philadelphia potter Jared Jaffe works almost exclusively with the teapot form. The primary reason is "the great possibilities for artistic expression within the limitations set by its function," he says. "I feel that function has to be considered along with the physical qualities of a teapot. It is the most integral aspect of the form, a fluid combination of handle, body, spout, lid and foot."

For organic and animalistic undertones, his "handles and spouts may become stems, horns, tails, snouts and so forth. The body of the teapot may take on a life of its own, transforming into snails, shells or tree trunks."

The construction technique is whichever method he feels suits the form best. Simple cones, spheres or cylinders are thrown on the wheel. The handles and spouts are usually pulled solid, and any unusual shapes are handbuilt or directly molded from an actual object.

Jaffe then makes plaster molds of the teapot for eventual slip casting. This step eliminates any cracking or

"Burnt Wood Teapot," 7 inches in height, slip cast and handbuilt, brushed with color variations of the same low-fire glaze.

separating of the various components. It also renders the teapot lightweight and of uniform thickness. When the teapot is removed from the mold, the surface is textured by carving. This allows the small undercuts that would otherwise be impossible to achieve from slip casting.

Glazing is with variations of the following base glaze:

Clear Gloss Glaze
(Cone 06–04)

Gerstley Borate	65%
Edgar Plastic Kaolin	10
Flint (325 mesh)	25
	100%

Colors are achieved with the following additions (all can be opacified by adding 5%–10% Superpax):

Deep Blue	
Cobalt Carbonate	2%
Bright Yellow	
Mason Stain 6485	10%
Green	
Copper Carbonate	5%
Light Purple	
Mason Stain 6006	10%
Orange	
Mason Stain 6121	10%
Dark Brown	
Red Iron Oxide	15%

Viewers at the "Art Now" exhibition at the Main Line Art Center in Haverford, Pennsylvania, were sometimes surprised to discover that the finished teapots are intended for use. Jaffe wants his work to enrich and enliven the familiar routines that people perform on a daily basis. He challenges himself to arrange seemingly disparate elements into a visually pleasing and coherent form that functions both as a teapot and as an artistic object. ▲

"Snailing on a Pea of Green Teapot," 8 inches in height, slip-cast and handbuilt whiteware, by Jared Jaffe, Philadelphia.

SCULPTURAL WORK—THE SIGNIFICANCE OF THE SURFACE

Many artists who create sculptural work prefer to do their firing in the electric kiln. For some potters, making the pots is akin to preparing a canvas, and the real art happens during the firing when the smoke and flames get a chance to paint the pots. When a sculptor may work for a week or more to create a single piece, it is not so tempting to consign it to the flames and see what happens. For ceramic sculptors, using an electric kiln is often about keeping control over the work, creating just the right surface, and preventing firing mishaps. An electric kiln can be brought up to temperature slowly and evenly to prevent stress and breakage. The neutral atmosphere reduces the unpredictability of glazing, and there is no ash to settle in just the wrong spot on a carefully finished sculpture. Furthermore, most types of work can be fired multiple times if necessary, especially if the temperature is lower with each firing.

Many sculptors who fire in electric kilns rely on texture to enhance the surface of their work. A highly textured surface, of the sort that usually isn't appropriate for functional pottery, can be used to add both visual and tactile appeal to a sculptural work. This texture can be created in the making process, by impressing, incising, paddling or scraping, and simply enhanced with stains and engobes, minimizing the use of glaze. In other cases, the texture is created by means of dry or crawling glazes.

Other sculptors focus not so much on the texture of the surface as on the color. These artists rely on a variety of approaches—some using glazes colored with commercial stains, others using glazes and underglazes, china paints, and lusters, and non-fired finishes—to achieve the desired effects. Where a pitcher meant for daily use must have a durable, fired-on finish, a piece destined for permanent display in a vitrine or on a mantelpiece can take advantage of any material the artist can think of to use, including acrylic paints, gold leaf, wax, and so on.

Another common characteristic of ceramic art pieces is the tendency to choose a firing temperature based on the desired characteristic, rather than on the formal maturing temperature of the clay. A potter who fires stoneware clay at earthenware temperatures must carefully fit the glaze to the clay to prevent crazing and porosity in the ware. A sculptor is freed from such concern, unless the work is to be situated outdoors in an inclement climate, and many choose to fire below the maturing temperature of the clay to reduce the stress of the firing. Electric kilns provide an easy way to gauge the exact temperature of the firing, and to duplicate desired results.

This chapter features sculptural and non-functional vessel-oriented work by several artists with very different styles. The unifying factor in all this work is not just the type of kiln used, but the fact that all the artists are highly concerned with the surface of their work. The work includes delicately sculpted porcelain, rough-textured stoneware, and assemblages consisting of multiple elements. Some of these artists incorporate personal and symbolic imagery in their work, while others comment on the nature of their chosen material by basing their work on the vessel form. They employ a great variety of materials in the making and especially in the finishing of their work, using the full range of expression of which clay fired in an electric kiln is capable.

Deeply Rooted

by Linda Ganstrom

"Deeply Rooted Divers," to 26 inches (66 centimeters) in length, handbuilt white stoneware with stains, fired to Cone 6.

Based on experiences in the fertile prairies of the central United States, my figurative sculptures are intended to evoke the spiritual connections I feel between the land and those who cultivate it. Growing up amid wheat, corn and cattle on a Kansas farm, I witnessed the annual cycles of planting and harvesting, and I developed a respect for and love of the environment. From the simple observation that crops require the energy of the sun and nutrients from the soil, I drew the analogy of the human soul's dependence upon external sources of nourishment. The products of culture—literature, theater, music and art—provide rich food for thought. Conversations on a soulful level of careful listening, earnest contemplation and honest response encourage an individual's growth. So too does quiet contemplation of the beauty and complexity of nature. Whether stimulating or withering, the environment one chooses to inhabit has a dramatic effect on the realization of one's potential.

This environmental effect is expressed in my sculpture series, "Deeply Rooted," through the connection of the vertical figures to their horizontal counterparts, as though one were literally growing upward from the other. The roots that bind the figures together symbolize a symbiotic relationship, as the figures rely upon one another for life. The lower, recumbent figures offer stability, rich resources and inner strength. The upper, more active figures branch out, taking risks that are made possible by the support of their partners.

The interdependence that I wish to convey through my work is more than the physical relationship between a plant and the soil; there is also a less tangible and less easily articulated connection. Living near Fort Hays State University, I am surrounded by an environment of prairie, canyons and chalk formations that contain evidence, in the plentiful crustaceous fossils, of the ancient seabeds that once covered Kansas. Inspired by these fossilized remains, I developed textures to cover my figures in a veil of mystery, removing them from the world of the easily understood. The surfaces evoke natural stone formations, fossils, feathers, scales, bark and prairie vegetation, but they can never be definitively identified. The viewer is encouraged to engage his or her imagination, drawing upon personal experiences to resolve the mystery of the imagery. There are no right or wrong answers; ultimately, the solutions one discovers when confronted by these kinds of enigmas connect to a larger picture of what one finds meaningful in life.

This realization led me, as a graduate student in the 1980s, to begin a series I titled "Seed Sisters." My immediate goal in these works was to explore the creative power of women in order to understand my own contribution to a larger community. I chose to investigate the metaphor of woman as vessel, carrying love, the seeds of new life and a host of ideas, emotions and dreams. I meant to refer to both the unprecedented potential of contemporary women and the more timeless potential of a universal woman. She loves, bears children, protects, heals, grieves, nurtures, works and worships. She is at once ancient and new.

Like my subsequent sculptures, the "Seed Sisters" are produced as slab constructions, but with particular attention given to building up layers. The surfaces are accumulations of paddled, scraped and scratched clay that are intended to reference the manner in which we form our characters, building upon layers of experience, knowledge and perception. I normally begin with a central column, then reveal the figure through a gentle pushing and pulling of the clay. The seed-vessel base often serves as a canvas for carvings and a layering of images that relate to the regulating concept. Daughter, mothers and lovers coalesced here, giving insight into the feminine mind and spirit.

To enrich the surfaces and to enhance the sense of individuality, I employ a wide variety of surface and firing techniques. Some figures are formed

"Deeply Rooted Rider," 30 inches (76 centimeters) in height, handbuilt stoneware, with slips, engobes and stains, by Linda Ganstrom, Hays, Kansas.

PHOTOS: AL SURRATT AND SHELDON GANSTROM

"Deeply Rooted Seed Sister—Eldest," 23 inches (58 centimeters) in length, handbuilt white stoneware with stains, fired to Cone 04.

from 10- to 12-inch sections of firm slabs that are stretched and scraped smooth. When leather hard, they are carved. After the bisque firing, they are painted with underglazes.

To create a more mysterious surface, I attach a thin slab to the smoothly scraped skin of the piece. These slabs are sliced from a solid block of clay with a cutoff wire, then impressed with ceramic stamps, shells and fossils. Next, they are stretched to a thickness of ⅛ inch by tossing them on a canvas-covered surface.

While waiting for the thinned slabs to dry somewhat, I roughly score the surface of the sculpture, then coat it with thick slip to remoisturize the leather-hard clay. Once the slabs are firm enough to be handled without destroying their textures, I score and slip their undersides, before attaching them to the sculpture. Next, I wrap the piece in plastic for at least a week to allow the moisture content to even out, then let it dry slowly and thoroughly before the bisque firing.

I use very thin glaze washes or stains to color and bring out the contrasts in these subtly textured surfaces. The stains are developed from a clear glaze. I add oxides or stains, then thin the glaze to about the consistency of skim milk. Experimentation is necessary to achieve the desired color and sheen.

Typical of the clear base recipes I use is this volcanic ash glaze that fires and matures anywhere from Cone 04 to Cone 9:

Volcanic Ash Clear
(Cone 04–9)

Volcanic Ash (from Lincoln County, Kansas)	60%
Gerstley Borate	40
	100%
Add: Bentonite	2%

Black Stain/Wash
(Cone 04–9)

Copper Carbonate	1 cup
Volcanic Ash Glaze	1
	2 cups

Golden Brown Stain/Wash
(Cone 04–9)

Milled Rutile	2 tablespoons
Red Iron Oxide	1
Volcanic Ash Glaze	4
	7 tablespoons

Blue Green Stain/Wash
(Cone 04–9)

Alumina Oxide	1½ tablespoons
Chromium Oxide	1
Cobalt Oxide	2
Volcanic Ash Glaze	16
	20½ tablespoons

I usually cover the piece in a copper stain, then sponge most of it off except for recessed areas. This will produce a

"Deeply Rooted Night Mother," 48 inches (122 centimeters) in height, grogged white stoneware, with stains, fired to Cone 04.

black, almost metallic shadow. Next, I layer on thin washes of red iron oxide, rutile or cobalt mixtures, rubbing them into the surface with a stiff paper towel. Finally, I highlight the piece with a rutile wash that turns a warm golden brown. To avoid heavy metal poisoning, I wear protective gloves and a face mask when working with the stains.

The sculpture is fired in oxidation to Cone 04, 6 or 10, depending on its final destination. Some works are surfaced with low-fire slips and engobes, fired, then layered with the stains.

While reading about pre-Columbian life and art before a 1991 trip to the Yucatan in Mexico, I happened across a reference to an ancient philosophy, which held that peace of mind could be achieved through acknowledging the different pulls of personality. For example, the desire for security is countered by a love of adventure. I decided to personify these pulls of personality as female ancestral figures in another series. These ancestral figures seemed to me more universal than the word grandmother would suggest, so I described them as the "Great Mothers."

As my work has progressed from the "Seed Sisters," which are metaphors for woman's potential, to the "Great Mothers," which represent a broader picture of my own personality and its projection onto the cosmos, I have endeavored to make sculptures that address the mysteries of life, that invite the viewer to examine him- or herself without judgment in order to bring about greater self-awareness. ▲

Suggestive Symbols

by David Benge

I create ceramic art of an eclectic nature, incorporating images and symbols from other cultures. Each has a specific meaning and purpose that is designed to affect the viewer in a unique way. For example, the ancient Egyptian winged scarabs are intended to suggest transformation and rebirth, while serpents have the effect of drawing the viewer out, in a sense forcing the observer to react. I must admit, however, that this can work both ways; some people are fascinated by serpents, while others are repulsed and pushed away. I don't worry about this. I view serpents in a classical sense. They have always been used in art as a symbol of metamorphosis, mystery, magic and sexuality.

Another ancient Egyptian symbol I also use is the ankh, which is shaped like a cross with a loop at the top and sometimes with a loop on each side under the cross arms. It is a symbol of life-force. It represents knowing ourselves on deep levels, a symbol of personal unity.

My representations of Earth are meant to be a symbol of wholeness. It's my belief that the world has the potential to come together with an entirely new kind of unity that would heal the current schism between the traditional mind and the higher self, and remind us of our true connection to the planet on which we live.

I've been involved in clay since 1965, when I studied ceramics with F. Carlton Ball and Susan Peterson. They taught me a lot about the technical side of ceramics-that it's only through a sound understanding and control of the materials that an artist can expect his or her true vision to take form.

"Dolphin World," 11½ inches in height, slip-cast and press-molded porcelaneous stoneware.

For ten years, I was a Cone 10 reduction potter in Santa Fe, New Mexico, then I returned to my native Los Angeles to work in various ceramics factories in the area. I did research and development in glazes and clay bodies, as well as designs for tableware and other porcelain products. These designs are still available on the market today.

In 1990, I struck out on my own to create artwork that would reflect my interest in metaphysics and personal transformation. I wanted to make art that would reach people on a more personal level, to make them think and feel in a new way.

My current body of work consists of slip-cast, press-molded porcelaneous stoneware fired to Cone 6 in an electric kiln. The clay is colored with stains and raw oxides; glazes are simple. In fact, most of these pieces are unglazed or utilize minimal glaze only in specific areas of the piece. I like simplicity and understatement.

For the most part, this work is produced as separate elements, which are assembled with epoxy. I create my own plaster molds, sculpting the original shapes out of Plasteline, then casting press or slip molds.

Sometimes, I paint backgrounds for these forms on gessoed wood panels, using oils, acrylics or casein. This allows me to complete an idea, which otherwise could not be done, given the chromatic limitations that exist with fired clay and glaze.

Having worked with clay for many years, I've learned that there is a natural evolution to the work. To keep it fun, you have to keep growing as an artist. Creativity is about fun. Anything less does not satisfy and eventually leads to boredom.

I see the whole process of making art as a challenge. When I get stuck or frustration sets in, I always remember my primary goal is to achieve some semblance of aesthetic beauty. People need beauty in their lives, especially now, and I try hard to supply that in some small way. ▲

PHOTOS: BILL ARON, DAVID BENGE

"Fire Scarab," 16 inches in diameter, glazed and lustered stoneware on oil-painted wood, by David Benge, Los Angeles, California.

Recipes

David Benge Stoneware Body
(Cone 6)

Alumina Oxide (Calcined)	1.0 parts
Wollastonite W-10	5.0
Nepheline Syenite (400 mesh)	12.0
Primas Feldspar	10.0
Edgar Plastic Kaolin	15.0
Kentucky Ball Clay (OM 4)	27.5
Kentucky Special	7.5
Pyrax Clay.	5.0
Sno Cal Kaolin	5.0
Flint (200 mesh)	23.0
	111.0 parts

Age for 30 days if using as a plastic body. Colored bodies are produced with a variety of stains and oxides:

Blue-white:
Harshaw D-355 Periwinkle Blue Stain......... 2.0 parts

Blue-purple:
Mason 6385 Purple Stain........ 7.5 parts

Black:
Harshaw Z-560 Black Stain..... 7.5 parts

Brown:
Ferro C834 Peach Stain......... 7.5 parts
Red Iron Oxide.................. 2.0 parts

Golden adobe:
Mason 6464 Zirconium Yellow Stain .... 4.0 parts
Red Iron Oxide.................. 1.0 parts
Rutile .... 2.0 parts

Dark cobalt:
Ferro C202 Blue Stain .... 5.0 parts

David Benge Matt Glaze
(Cone 6)

Whiting	10 %
Zinc Oxide	20
Frit CC-257 (Ferro)	10
Primas Feldspar.	35
Kentucky Ball Clay (OM 4)	10
Flint (325 mesh)	15
	100%

Color variations are achieved with the following additions:

Black:
Cobalt Carbonate 2%
Copper Carbonate 4%
Drakenfeld 4346 Red Brown Stain 3%

Periwinkle blue:
Harshaw D-355 Periwinkle Blue. 4%
Zircopax 7%

Note: The matt glaze must be deflocculated. Slake glaze with 60 cubic centimeters of water for every 100 grams dry glaze. Let sit overnight, then mix as you add drops of Darvan that has been diluted with 50% water. Add only as many drops as you need to thin the glaze, usually 12-20 drops for a 3000-gram batch.

David Benge Gloss Glaze
(Cone 6)

Whiting	15.00%
Primas Feldspar.	23.00
Frit 400 (Hommel)	5.00
Frit 3292 (Ferro)	18.00
Edgar Plastic Kaolin	12.00
Flint (325 mesh)	27.00
	100.00%
Add: Bentonite	3.40%

Ball mill for seven hours for a smooth melt. Color variations are achieved with the following additions:

Opaque white:
Zircopax 10.00%

Gray:
Harshaw Z-560 Black Stain 0.25%

Rectangular platter, 13 inches (33 centimeters) in length, slab-built and incised stoneware, with multiple slips, fired to Cone 6.

Rectangular platter, 14 inches (36 centimeters) in length, slab-built and incised stoneware, with multiple slips, fired to Cone 6.

Mary Barringer's Conversations

by Leigh Taylor Mickelson

Her pots are quiet, but not silent. They begin whispering to you the moment you come into contact with them. If you listen, they will start a conversation with you about nourishment, space and celebration. Behind the words will be echoes of history, landscape and culture. At Mary Barringer's solo show at Baltimore Clayworks, I felt an instant connection to her pots. It took some much-valued time with the work, as well as with the maker, to understand why.

The way ceramics artists first come to clay seems to feed the core of their visual language. Barringer was led to clay in a roundabout way. She was studying sculpture at Bennington College in the early 1970s, making figurative work at a time when her instructors and the greater "art world" thought there was not much that could be done with the figure that had not already been done before (sound familiar?). At about the same time, she had signed up for a required class: ceramics. The combination of the responsiveness of the material and the nonresponsiveness of her teachers led her away from the "hierarchies" of the art world and into a more personal journey with clay.

Although Barringer started out making wheel-thrown stoneware pots, she soon detoured from the wheel toward handbuilding, which opened up huge possibilities that she had never considered before. "There were so many pots made for different purposes and cultures, made by people with no written record, so pots were their record, their stories. This was very vibrant for me." From this point on, Barringer's work took on a sense of history and background, suggesting the passage of time through layering of textures and colors, faint nuances to discover as the work was handled. It was at this point that the conversations began.

In 1988, after 16 years of selling her pottery and sculpture out of various studio spaces, she moved to a small town in western Massachusetts. Here, the landscape—formed by glaciers and water—seems to "echo what's going on internally" as Barringer works. "The process of forming landscape is like handbuilding pots—there's scraping and burnishing; rocks bore holes into other rocks. Like pots, rocks are things that are formed incrementally over time."

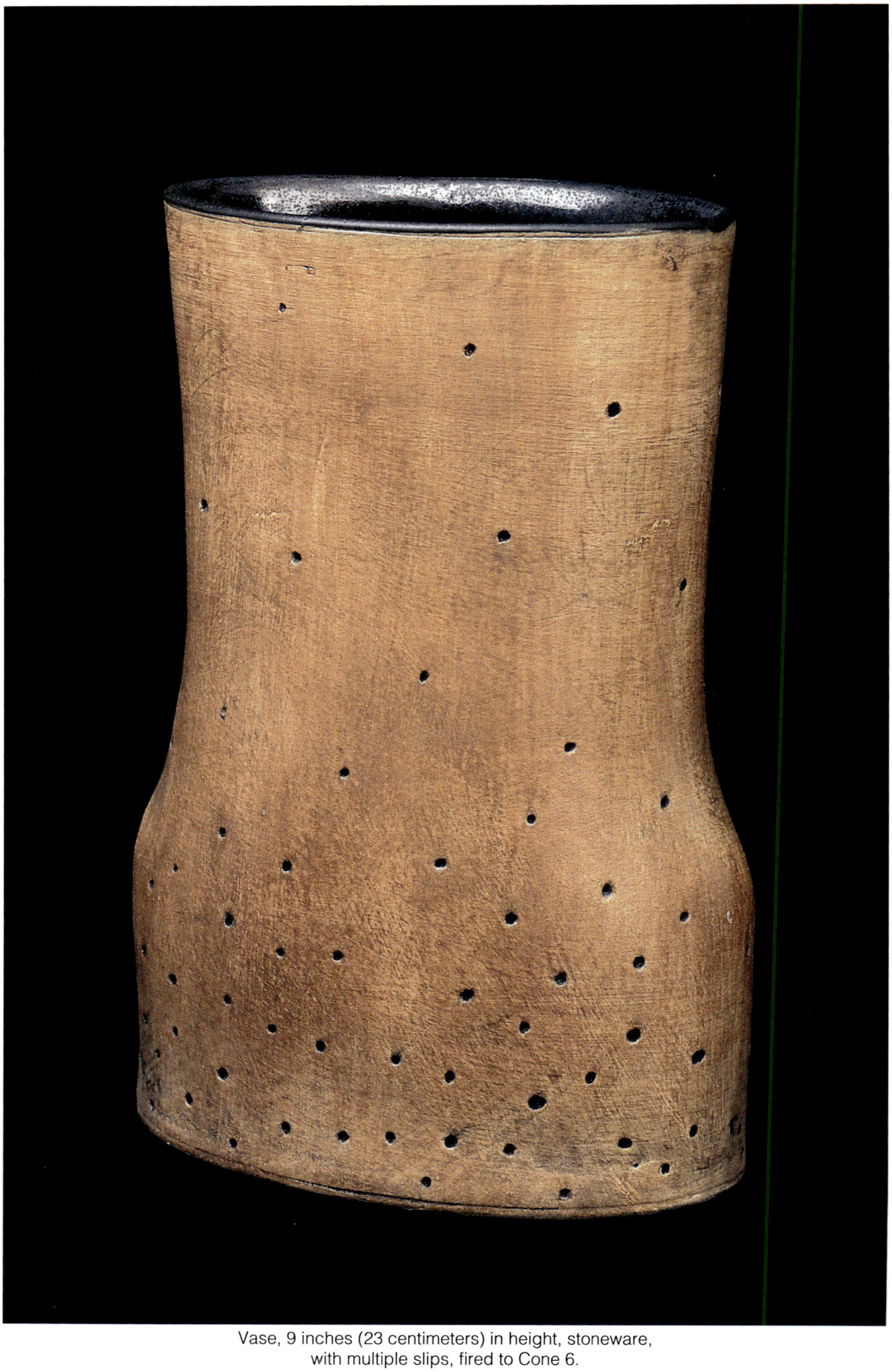

Vase, 9 inches (23 centimeters) in height, stoneware, with multiple slips, fired to Cone 6.

Ovoid dish, 12 inches (30 centimeters) in length, slab-built and incised stoneware, with multiple slips, fired to Cone 6.

For a while, Barringer focused primarily on making sculptural forms, but then she began to miss making pots for use. She missed the "more informal, personal set of experiences that go along with domestic space." So the pieces shown at Baltimore Clayworks in "New Work from a New Place" consisted of functional vessels: platters, teapots, vases, plates, creamers and basins.

Within this work, there is a balance, or rather a tension, between form and function, texture and line, color and image, hard and soft, warm and cool. Although fired to Cone 6 in an electric kiln, the surfaces have a richness and depth that I had thought could only be achieved in a reduction atmosphere.

Barringer creates a texture—lines, scratches, punctures, grooves—with her many tools. Then she brushes on some slip, scrapes it off, brushes on another slip, wipes it off with a green scrub pad, over and over. And once the pot has been bisque fired, she begins again, layering on various thicknesses of slips and glazes, then scraping or wiping them off. Here, especially, her technique mirrors the weathering of the New England landscape—the rhythmic motions and fortuitous scraping of making and erasing marks.

These marks are an essential part of the conversation that her pots have with their viewers. For instance, two rectangular platters were hung next to each other; though they were the same shape and size, they were very different in feel because of Barringer's use of texture and color. The first was a warm hazelnut brown with strong parallel lines and lighter, seemingly random, diagonal scratches that suggested weathering or passage of time etched into the surface. Embedded in the lines was a dark brown

Creamer, 4Z\x inches (11 centimeters) in height, slab-built stoneware, with multiple slips, fired to Cone 6.

Teapot, 6Z\x inches (17 centimeters) in height, handbuilt stoneware, with slip, fired to Cone 6, by Mary Barringer, Shelburne Falls, Massachusetts.

slip. Under a strip of green glaze on the bottom was some writing. At first, I thought, "Oh, this will tell me what this plate is about...," but then I realized that I could only pick out a word or two, "cold day...sky above...."

On the second platter, the background was a soft cool blue, marked with deep black grooves that were uniform yet randomly placed throughout. On top of the light blue background was a darker blue form—two structures rising out of a thick red glossy strip at the bottom. The image almost looked like a worn building, one that was falling in on itself, or perhaps two torsos, leaning in toward each other to ward off the surrounding cool air.

Placed throughout the gallery were seven vases, all torsos in their own right. Slab built, they had quite apparent hips and belly buttons, spots of bright glossy red glaze that suggested the core of the pot, the center of the pot's being.

Barringer incorporates this red dot into a number of her pieces—plates, creamers, dishes and basins—drawing the eye out of the thick layers of slip and texture to this poignant, potent splash of color. The red gives us something to focus on—a spot of color within the rich earth tones, a familiar place to see and touch as you hold the piece, rotating it in your hands. It speaks of the kind of familiarity that relates comfort, like your favorite coffee mug in the morning or those hand-knit wool socks that you wear on cold days.

There is no doubt that Barringer is "fascinated by the conversation that goes on between people and things—and between people through things—and I want to participate in that with my work. It means engaging in ideas that are not just personal, and that my personal exploration goes on against the backdrop of a larger, cultural set of ideas about objects and functions. I like that my pots go out and lead a life separate from me, and become a part of someone else's life. That potential is a huge gift for a potter." ▲

White Slip Base
(Cone 6)

Frit 3124 (Ferro)	10.0 %
Nepheline Syenite	15.0
Ball Clay	25.0
Edgar Plastic Kaolin	25.0
Flint	25.0
	100.0 %

Color variations are mixed with the following additions:

Black:

Black Stain	10.0 %
Red Iron Oxide	8.0 %

Blue-Black:

Black Stain	10.0 %
Cobalt Carbonate	2.0 %

Gray-Green:

Chrome	3.0 %
Copper Carbonate	3.0 %

Light Green:

Copper Carbonate	5.0 %

Strong Green:

Chrome	6.0 %

Blue-Green:

Chrome	3.0 %
Cobalt Carbonate	1.5 %

Teal:

Copper Carbonate	3.0 %
Teal Stain	6.0 %

Medium Blue:

Cobalt Carbonate	1.0 %
Rutile	3.0 %

Cream:

Rutile	5.0 %

Yellow:

Yellow Stain	10.0 %

Pink:

Pink Stain	10.0 %

Vitreous Black Slip
(Cone 6)

Borax	10.0 %
Nepheline Syenite	23.0
Kaolin	22.0
Ball Clay	23.0
Flint	22.0
	100.0 %
Add: Cobalt Oxide	1.0%
Copper Oxide	4.5%
Red Iron Oxide	4.5%

Redart Slip
(Cone 6)

Lithium Carbonate	10 %
Spodumene	10
Cedar Heights Redart	80
	100 %
Add: Red Iron Oxide	2%

Ash Engobe
(Cone 6)

Talc	15 %
Wood Ash (unwashed)	20
Cornwall Stone	25
Frit 3124 (Ferro)	15
Ball Clay	25
	100 %

For a white/tan variation, add 7.5% tin oxide; for pink/brown, add 1.5% nickel oxide and 8% pink stain.

Stevenson Vitreous Engobe
(Cone 6)

Whiting	3.33 %
Feldspar	8.88
Frit 3124 (Ferro)	50.00
Ball Clay	11.11
Kaolin	13.34
Flint	13.34
	100.00 %
Add: Tin Oxide	11.11%

Single-Fire Ash Glaze
(Cone 6)

Wood Ash	19 %
Potash Feldspar	25
Kaolin	12
Red Clay	13
Flint	31
	100 %
Add: Cobalt Carbonate	1–5%
Red Iron Oxide	15%

Burlington Base Glaze
(Cone 6)

Lithium Carbonate	2.03%
Strontium Carbonate	5.39
Whiting	12.21
Soda Feldspar	20.35
Albany Slip	22.38
Ball Clay	25.43
Zircopax	12.21
	100.00%

Color variations are possible with oxide, carbonate and/or stain additions.

Chinese Ash Slip
(Cone 6)

Gerstley Borate	2.92 %
Wood Ash	48.54
Ball Clay	48.54
	100.0 %

RC Slip
(Cone 6)

Whiting	30.0%
Kona F-4 Feldspar	20.0
Cedar Heights Redart	40.0
Flint	10.0
	100.0 %

Glassy Slip
(Cone 6)

Gerstley Borate	5 %
Lithium Carbonate	80
Bentonite	15
	100 %

Dense White Engobe
(Cone 6)

Frit 3110 (Ferro)	10 %
Nepheline Syenite	20
Ball Clay	30
Zircopax	40
	100 %
Add: Bentonite	2%
Macaloid	1%

Yellow Oxide Glaze
(Cone 6)

Lithium Carbonate	4.35 %
Zinc Oxide	18.48
Frit 3110 (Ferro)	28.26
Kaolin	31.52
Flint	17.39
	100.00 %
Add: Tin Oxide	4.35%
Copper Carbonate	0.44%
Iron Oxide	4.35%
Yellow Ocher	0.54%

Metallic Black Glaze
(Cone 6)

Gerstley Borate	10.75 %
Whiting	5.48
Feldspar	78.86
Kaolin	4.91
	100.00 %
Add: Cobalt Oxide	2.07%
Copper Carbonate	4.14%
Manganese Dioxide	4.14%

Super-Dry Matt Glaze
(Cone 6)

Gerstley Borate	5 %
Lithium Carbonate	6
Strontium Carbonate	15
Nepheline Syenite	60
Calcined Clay	8
Flint	6
	100 %

Color variations are possible with oxide, carbonate and/or stain additions.

Sheen-O Glaze
(Cone 6)

Soda Ash	2.9 %
Gerstley Borate	4.9
Spodumene	22.8
Nepheline Syenite	54.5
Ball Clay	14.9
	100.0 %
Add: Chrome	0.5 %

"Forbidden Flower," 12 inches in diameter, handbuilt porcelain, fired to Cone 5.

Keisuke Mizuno

Porcelain sculpture, by St. Cloud, Minnesota, artist Keisuke Mizuno, titled "Forbidden Fruits" and "Forbidden Flowers" are "about the beauty of inseparable dependence, and visual disparity between life and death.

"Much of my work depicts slugs thriving on a fruit or flower, which contains a miniature skull inside," Mizuno explains. "Fruit and flowers are used to create a familiar and intimate environment for the audience. They also symbolize fertility and impermanence of life. The transient nature of the fruit or flower is contrasted by the slow movement of slugs, which aggressively eat away a fruit and flower, exposing the miniature skull embedded as a seed."

The small, intricately detailed forms are handbuilt from a porcelain body softened in color with the addition of a yellow stain:

Porcelain Body
(Cone 5)

Nepheline Syenite	26.00%
Grolleg Kaolin	58.00
Flint	16.00
	100.00%
Add: Macaloid	2.00%
Yellow Stain	0.25–1.00%

After a bisque firing, commercial underglazes and colored glazes are applied, then fired to Cone 5.

Clear Glaze
(Cone 5)

Gerstley Borate	8.00%
Talc	2.00
Whiting	7.00
Zinc	5.00
Kona F-4 Feldspar	48.00
Kaolin	9.00
Flint	21.00
	100.00%

Color variations are produced with the following additions:

Yellow:

Yellow Stain	1.00–5.00 %

Green:

Copper Carbonate	1.00–5.00 %

Yellow Green:

Copper Carbonate	0.25–2.00 %
Yellow Stain	5.00 %

Orange

Degussa Orange Stain	0.25–3.00 %

Yellow Orange:

Degussa Orange Stain	0.25–3.00 %
Yellow Stain	5.00 %

Red

Degussa Red Stain	0.20–7.00 %

Finally, china paints are used to "create subtle color change throughout." The china paint applications/firings are repeated three to five times. ▲

PHOTOS: DAN DANNEHY, CRAIG SMITH

"Forbidden Fruit," 5 inches in height, handbuilt porcelain with underglazes, glazes and china paints.

"Forbidden Fruit," 7 inches in height, handbuilt porcelain with underglazes, glazes and china paints, by Keisuke Mizuno, St. Cloud, Minnesota.

JoAnn Schnabel

by Rachel Marie-Crane Williams

After nearly a decade as head of the ceramics department at the University of Northern Iowa, JoAnn Schnabel had the opportunity to spend a few months away from her teaching responsibilities to concentrate on creating new work. Several weeks spent at Penland School of Crafts in the mountains of North Carolina contributed not only time and space but also energy to the sabbatical experience.

In late April, chairs caked in slip, dusty boxes, and racks of pots in various stages of production lined the walls of Schnabel's studio at Penland. Eric Clapton music filled the cool, moist air. On a table was a series of small demitasses glazed in purple. A finished set of teapots with exaggerated spouts and handles sat nearby; pushed together in a mass, they formed a colony of whimsy. "These pots," she explained, "fulfill a playful need in my art making."

Those quirky teapots, ewers and demitasses were so different from anything she had made in the past. Previously, her work had focused on large sculptural forms and tiles inspired by her spectacular garden. Assembled from parts, the new vessel forms reflected her interest in the play of geometry.

"These vessels are more of a game. I ask myself, 'Where do I want parts, and how am I going to put those parts together—the different textures of the carved forms, the different shapes? How is color going to unify or separate things? How do I want things seen?' If something looks too disjointed, I either unify

"Tea for Who?" 15 inches (38 centimeters) in height, assembled terra cotta, fired to Cone 3.

it or make it really disjointed with color. Subtle colors can create mood."

In addition to the vessels, there were two large abstract forms that looked somewhat like stylized mountains or waves firmly planted on a table littered with stamps, fruits, and odds and ends. Between the solemn sculptural forms and the animated vessels runs a common thread—Schnabel's thoughts about nature, domestic rituals and the process involved in making.

About the vessels, she commented: "There is something kind of funny—I am looking at ewers and thinking about a circus. I am relating the shapes to things more playful. For example, I look at a spout and relate it to an elephant trunk. There is a quality in the pots that is playful; I don't want the sculpture to go there. One feeds the other. Having both to work on is a nice balance for me. The sculptural things are not preplanned, like the vessels. I am trying to work more intuitively.

"The vessels are lighthearted; they have a format....The sculptures have no preconceived format. They are more contemplative....Some things that are considered in the sculpture naturally carry through into the pots....It's just a different play of elements."

Pointing to the sculptures, she observed: "These forms are very time consuming. They are...solid with weight....I am envisioning more sitting on a floor, so that you could interact with them in a different way than you would interact with the pots. These forms embody the experience of visual and physical weight. They look more somber."

Inspiration for the sculptures came from the quiet she found on long walks in the woods. The act of walking also played an important role in their perceived placement. "I am really thinking about how those forms will go across the gallery floor....I want them to split the floor, so viewers will have to have personal interaction with these pieces. People will have to walk around, or follow the space between them like a path. I want them to think about the micro- and macrolandscape, the rocks in the path."

There were also several small podlike forms in Schnabel's studio. Each was carved, some defined by sinewy ridges, others by squares of clay touching on the corners. Each made some sort of noise when disturbed, as a sphere trapped inside rattled around playfully.

Schnabel started this series of pods as part of a daily ritual to unwind after teaching: "I was very frustrated trying to teach and make work at the same time. A couple of years ago, I started carving these solid forms. The form would emerge from clasping my hands together while holding a lump of clay. I liked making these things that fit into my hands. They gave me the feeling of accomplishment, of getting something done. I could spend a few hours carving and have something finished."

The desire to tempt people to touch these forms is part of her involvement as their maker. "The tactile quality, when I make something, is important. That is part of the end experience that I want for the viewer."

The pods are displayed on wall-mounted shelves to encourage interaction with the viewer. "I am influenced by a lot of different things. How am I going to display this, and how is the space going to be arranged? These

PHOTOS: TOM MILLS, JOANN SCHNABEL, RACHEL WILLIAMS

"Repository Study I," to 18 inches (46 centimeters) in height, handbuilt terra cotta, with brushed glazes, fired to Cone 3.

Schnabel glazing pod forms in her Cedar Falls, Iowa, studio.

"Opposites Attract II," 19½ inches (50 centimeters) in length, glazed terra cotta, fired to Cone 3 in oxidation.

questions are part of wanting them [the pods] to have a special place—to live in a certain spot—rather than sitting on an isolated pedestal."

For Schnabel, there always seems to be an internal dialogue that fills her studio and informs the choices she makes about her work; this dialogue is heavily influenced by her work with plants, discussions with other artists, travel and her walks in the woods.

She enjoys working in a communal environment. Her studio at Penland was a large space shared with several other artists. Being away from home, "divorces you from your everyday life," she says. "There are no committee meetings, I don't have to return business calls and there is no house to clean; this allows me to be in the studio ten hours a day."

In 1999, she received a Professional Development Grant from the Iowa Arts Council to travel to Spain. "I went to Barcelona for six days as part of the Ceramic Caravan Tour offered in conjunction with the "Ceramic Millennium" symposium in Amsterdam. I took a side trip to Barcelona. Gaudi's architecture has been a huge influence on my work, and seeing his buildings in person really struck me."

How Schnabel manipulates clay directly relates to the organic qualities of Gaudi's architectural forms. "When you are dealing with a medium that is process oriented and equipment intensive, there are all these other things that come into play. I was drawn to clay from the beginning. I tried other things, but came back to clay. There is something about taking a lump of nothing and making something out of it."

"Parts of Whole," 16 inches (41 centimeters) in height, terra cotta, with brushed glazes, fired to Cone 3 in oxidation.

Working with wet clay is perhaps Schnabel's favorite part of the process. Her least favorite is glazing, and firing is simply a means to an end. Most of her work is fired to Cone 3 in oxidation.

When beginning a piece, she does not really think about the final stages. Instead, she waits until after it is bisqued to consider color. Color is used to dissect the original form. Currently, her palette is festive; it features rich purples, aquas, and earthy yellows and reds.

The various layers of glazes are recorded in a sketchbook, so that she can remember how each combination was achieved. She also records her thoughts and images about emerging pieces. "The sketching I do before I start working is a beginning point. The pieces don't end up as [products of] the sketches. I deviate.... I can always come back to the sketch. I work with what is in front of me at that moment.

"Being a one-person area at the University of Northern Iowa, I have found it really hard to make meaningful work during the school year. The reality is that the majority of my work is made in the summer when I can be in the studio all day."

In addition to being a teacher and an artist, she is also an avid gardener. "There is something about the phenomenon of life cycles that somehow influences my work. And seeing things as they are growing and changing..., seeing how plants look as they are coming out of the ground and are in full bloom..., I think that naturally finds its way back to the studio."

Walking each day also has an impact on her work in the studio. Gaining an intimate awareness of a given environment by taking the same walk over and over again, and seeing the landscape change as the seasons change are inspiring to Schnabel. "The walk is a necessity—I have to take the dog out.... Walking gives me time to think about what I am working on....I bounce ideas around while walking."

Ideas reflecting interaction with the environment, process, time away from the demands of teaching, and her thoughts surrounding objects used in domestic rituals informed the work begun at Penland. Returning to the University of Northern Iowa and her gardens marked another transition.

Work completed during this leave period was presented in a solo exhibition at the Hearst Center for the Arts in Cedar Falls, Iowa. ▲

"Beehive Goblet," "Whirling Goblet" and "Rattle Goblet," to 11 inches (28 centimeters) in height, handbuilt terra cotta, with brushed glazes, fired to Cone 3 in oxidation, by JoAnn Schnabel, Cedar Falls, Iowa.

Recipes

JoAnn Schnabel builds all the tile pieces from the following Cone 04 sculpture clay, while sculpture and vessels are built from a Cone 3 clay body:

Cushing Outside Sculpture Clay
(Cone 04)

Wollastonite	5 lb
Frit 3124 (Ferro)	20
Nepheline Syenite	5
A. P. Green Fireclay	28
Barnard Clay	2
Cedar Heights Redart	30
Kentucky Ball Clay (OM 4)	10
Coarse Grog	15
Fine Grog	10
Medium Grog	5
	130 lb
Add: Red Iron Oxide	6 lb

Schnabel Terra Cotta Clay
(Cone 3)

Talc	10%
A. P. Green Fireclay	10
Cedar Heights Redart	50
Kentucky Ball Clay (OM 4)	10
Medium Grog	10
Silica Sand	10
	100%
Add: Red Iron Oxide	1 %
Bentonite	2 %

To each 100-pound batch, Schnabel also adds a small handful of nylon fibers.

Clear Satin Matt Glaze
(Cone 3)

Gerstley Borate	50%
Edgar Plastic Kaolin (EPK)	20
Flint	30
	100%

Color variations are possible with additions of various oxides, carbonates and/or stains; for a more opaque version, add 10% Zircopax.

Schnabel Satin Glaze
(Cone 3)

Whiting	20.5%
Frit 3134 (Ferro)	10.0
Nepheline Syenite	69.5
	100.0%
Add: Titanium Dioxide	5.5 %
Bentonite	2.0 %

A nice satin with good color response to variations of 8% commercial stain.

Schnabel Revised Satin Glaze
(Cone 3)

Whiting	15.86%
Frit 3124 (Ferro)	37.45
Nepheline Syenite	15.86
Edgar Plastic Kaolin (EPK)	18.68
Flint	12.15
	100.00%

A bright satin with good color response to commercial stains.

Fairdale Cream Glaze
(Cone 3)

Dolomite	11.63%
Gerstley Borate	11.63
Frit 3110 (Ferro)	27.91
Kona F-4 Feldspar	11.63
Spodumene	23.25
Kentucky Ball Clay (OM 4)	13.95
	100.00%
Add: Opax	13.95 %
Titanium Dioxide	2.33 %

A soft ivory satin glaze suitable for color variations with oxides and carbonates.

Granby Base Glaze
(Cone 3)

Gerstley Borate	11.36%
Frit 3110 (Ferro)	31.82
Kona F-4 Feldspar	43.18
Kentucky Ball Clay (OM 4)	13.64
	100.00%
Add: Opax	11.36 %
Tin Oxide	2.27 %

A shiny satin that works well with stains.

Oz Strontium Glaze
(Cone 3)

Dolomite	10.20%
Strontium Carbonate	32.66
Frit 3124 (Ferro)	26.53
Spodumene	22.45
Cedar Heights Goldart	8.16
	100.00%
Add: Titanium Dioxide	2.04 %

For a strong green, add 3% copper carbonate.

Glaze XXVI
(Cone 3)

Barium Carbonate	28.8%
Magnesium Carbonate	13.0
Whiting	8.1
Custer Feldspar	30.6
Flint	19.5
	100.0%
Add: Bentonite	2.0 %

A very dry surface; works well with an addition of 8% stain. **Not intended for use on surfaces that come into contact with food or beverages.**

The Totemic Sculptures of Ted Vogel

by Kate Bonansinga

We primeval forest felling,
We the rivers stemming, vexing we and piercing deep the mines within,
We the surface broad surveying, we the virgin soil upheaving,
Pioneers! O pioneers!

Walt Whitman, *Leaves of Grass,* 1855

Thus was the westward expansion described by a mid-19th-century American poet. Almost a century and a half later, visual artist Ted Vogel comments on this same phenomenon in what he calls "landscapes": totemic combinations of ceramic birds, cages, human heads, fish and tree stumps.

Whereas Whitman observed the settling of the West from Brooklyn, Philadelphia and Washington, D.C., Vogel has lived his entire life within the land he depicts. He spent his youth and undergraduate years in South Dakota; received an M.F.A. from the University of Colorado; lived in Helena, Montana, where he was the assistant director at the Archie Bray Foundation (1991–1994); and now resides in Portland, Oregon, where he is a professor of art at Lewis and Clark College. Vogel is uniquely positioned (in a way that Whitman certainly couldn't be) to comment upon the human impact on the region: he has the benefit of hindsight combined with firsthand knowledge of the complex interaction between humans and the land of the northwestern United States.

The stacked totems that are the centerpieces of Vogel's current body of work speak of his attachment to the land. For example, "Roost Stack" consists of five individual earthenware elements, each drilled vertically through its middle so that it may slip over the thin, steel rod or pin that anchors it to the top of the previous element. On top of the wide base is a tree stump, above that a fish, then an upside-down, gilded human head and, finally, a bird, which has come to be Vogel's signature form.

PHOTOS: BILL BACHHUBER

"Roost Stack," 21 inches in height, earthenware with terra sigillatas and gold leaf, stacked on steel rod.

This order is deliberate, referencing the inhabitants of water, earth and sky. The tree stump often serves as a foundation to Vogel's totemic narratives. It physically supports the other sculptural elements and conceptually refers to the tree as the support for many life forms: it provides shade for streams and fish; oxygen for mammals; homes for birds.

Vogel intends his artwork to be "subtly environmentalist," and in "Roost Stack" the stump gives the appearance of having been gnawed by a beaver as much as chopped by an axe. But many of his pieces include cleanly cut tree stumps that might be seen as a comment on intensive logging practice. And though Vogel means for his androgynous ceramic heads to be references to "humanness" and the human as a part of the natural world, they can also be interpreted as decapitated, symbolic of detachment, distant from previous relationships with the land.

In "Roost Stack," a black bird is perched on the head. "Many cultures believe that the bird wards off evil, offering protection from harm....We are continuously intrigued by birds, their ability to fly, their wildness, their freedom. Often we cage them, bringing a bit of the outside inside," Vogel writes.

One of the artist's important early aesthetic influences was his grandmother's collection of colorful, kitschy ceramic birds from the 1950s. She displayed these in her sunroom to impart the liveliness and freedom of the wilderness. But in Vogel's sculptures, it's the heads, not the birds, that are caged.

People are now confined to smaller and smaller spaces with dwindling resources. In turn, we are more dependent upon one another. After generations of celebrating the independence and self-reliance of the pioneers, both of these ideals have become almost impossible to maintain.

In "Cup Cage," a hand rests on a tree stump. It holds a cage in the form of a cup and within this cup is a small, gold head. If the hand suggests one human, he or she is holding and caring for another. He might also be taking something that doesn't belong to him. The totemic form of the sculpture is particularly appropriate to the latter interpretation, as totems are often associated with native peoples who, in the United States of Whitman's time, were subjected to the usurping of their land and livelihood by the pioneers.

Vogel's narrative landscapes began as literal landscapes when the artist was in graduate school in the early 1980s. His use of the land as his subject led to his interest in trees (in which he began to see the form of the human figure) and in birds. In 1992, Vogel began "River Keepers," his ongoing project of placing his ceramic birds in various places in the outdoors; on tree branches or stumps, amongst some rocks, on the ground. He calls these places "power sites" and each holds special meaning for him, often related to fly fishing. An unusual juxtaposition of public and private art, the birds are accessible to anyone who happens upon them, and yet they are, for the most part, in remote spots that few people visit.

Where Vogel's grandmother's collection of art brought the outdoors in, Vogel brings art to the outdoors. He informally solicits people's responses to these ceramic birds (he has placed about 100 of them) and considers these responses to be important to our understanding of the relationship between humans and birds.

All of Vogel's sculptures begin as a sketch. For his current body of work, these preliminary drawings are sketchbook size, but in the past, when he had access to a large studio, some of them were up to 30×40 inches. He would tack them to the wall and refer to them as he worked on the ceramic piece.

Most of the sculptural elements are pinched or coil built. A few are press molded or thrown and altered. Richly colored terra sigillata, glazes or stains cover the surfaces, which are occasionally finished with a "room-temperature glaze," or paint, after the final firing. Vogel also gilds some elements by applying gold leaf in the traditional manner: he sizes the surface, lets it dry until it's tacky, then applies squares of very thin gold.

The use of metallic coloring is particularly appropriate, as Vogel refers to his totemic sculptures as "trophies" that pay homage to birds. Again, this term encourages another interpretation; that is, "trophy" can refer to a stuffed animal or animal head.

These heads and birds are elements of Vogel's artwork that speak to the entire spectrum of the human experience. In balancing them, often seemingly precariously, with hands, cages, tree stumps and abstract forms, he seems to be recommending that people strike a similar balance of needs with responsibility to one another and to the environment. ▲

"Cup Cage," 18 inches in height, earthenware with terra sigillatas and gold leaf.

Recipes

White Sculpture Clay
(Cone 06–3)

Talc	20.0 lb
Bentonite	1.0
Cedar Heights Goldart	15.0
Hawthorn Bond	40.0
Kentucky Ball Clay (OM 4)	10.0
Coarse Grog	12.5
Medium Grog	10.5
Fine Grog	2.5
	111.5 lb

Works well for raku.

Ted's Black Stain

Black Copper Oxide	1 part
Mason Stain 6616	4
Gerstley Borate	4
	9 parts

If you leave out the black copper oxide, this recipe will produce a nice, slightly dry black at Cone 04–01; the higher you fire, the glossier it will be. For a slightly iridescent black, vary the copper amount.

Busch White Terra Sigillata

Borax	40 grams
Frit 3134 (Ferro)	200
Cedar Heights Goldart	1800
	2040 grams

Add to a mixture of 20 grams Calgon and 1 gallon water; ball-mill for 24 hours, then decant. For color variations, add 5%–10% stain or 0.5%–1.0% oxide; for a smoother finish, ball-mill the colored terra sigillata again, but do not decant again.

Weiser Terra Sigillata

Cedar Heights Goldart	4990 grams
Yellow Ocher or Yellow Iron	200
	5190 grams

Add to a mixture of 25 grams soda ash and 50 grams Calgon in 3.2 pounds water. Produces a nice orange brown at the Cone 06–04 range; yields a darker finish, when fired higher.

"Fish Cage," 23 inches in height, earthenware, with terra sigillatas and gold leaf, by Ted Vogel, Portland, Oregon.

Art Base Glaze
(Cone 06–04)

Barium Carbonate	28.65%
Lithium Carbonate	2.28
Whiting	9.01
Zinc Oxide	10.25
Custer Feldspar	34.35
Flint	15.46
	100.00%
Add: Copper Carbonate	2.85 %
Nickel Oxide	0.47 %
Rutile	2.85 %

Semitransparent matt green; for use on sculpture only.

Forbes Green Glaze
(Cone 06–04)

Frit 3124 (Ferro)	45%
Frit 3304 (Ferro)	45
Kaolin	10
	100%
Add: Copper Carbonate	5 %

Nice translucent green.

BM-3 Glaze
(Cone 06–04)

Barium Carbonate	40%
Nepheline Syenite	45
Spodumene	10
Flint	5
	100%
Add: Copper Oxide	4 %

Dry blue to deep purple; for use on sculpture only.

Soda Blue Glaze
(Cone 06–04)

Gerstley Borate	3500 grams
Soda Ash	500
Frit 3110 (Ferro)	3500
Kaolin	250
Flint	500
	8250 grams
Add: Copper Carbonate	150 grams

Produces a beautiful translucent water blue; also works well as a raku glaze.

A Personal Aesthetic

by Howard Koerth

PHOTOS: HOWARD KOERTH, STEVIE SCHEIDEMANTEL

"Ritual Vessel (43) I.93," 13 inches in height, wheel-thrown and altered earthenware, with dry lithium glaze, fired to Cone 010, by Howard Koerth, Weatherford, Oklahoma.

The vessel is a human object. To see a vessel, or even a shard, is to respond to an individual expression of the shared experience of our universal history as vessel makers and vessel users. The vessel is the record of human existence, the recognition of many cultures, the chronicle of our needs and desires. With the vessel, we simply function, extravagantly contain, decoratively hold or ritually petition for fulfillment. Through the creative use and interpretation of structure, form and surface, this object, like any object, can move beyond its literal definition and become the basis for visual expression.

My exploration of the vessel began in 1980. After a year of general studies in college, I decided to major in art (rather than always wish I had). The choice between sculpture and ceramics was made with romantic visions dancing in my head of being a mountaintop potter. As soon as the first vessel (a very simple vase that I still have) came off the wheel, I was hooked, or perhaps seduced would be a more appropriate

Two views (disassembled and assembled) of "Standing Ewer (Effigy Series)," 17 inches in height, glazed earthenware, fired to Cone 010.

word. To shape, form and pursue the vessel became the focus of my creative activity and development.

While completing my B.F.A. at the University of Kansas, I studied under David Vertacnik, whose background from Alfred University and personal work in sculpture had a profound effect on the development of my personal aesthetic. David was keenly aware of the history and formal concerns of the vessel. He also had a good eye for detail and the possibilities of form offered by this most traditional of ceramic formats. In addition, he had a sculptor's eye for the activity of form in space and a natural curiosity about, and need for, the individual use of materials, objects and forms for personal expression.

I soon discovered that I had little interest in the disciplined repetition required of a production potter. In retrospect, I recognize that it was exploration and examination that fascinated me, not design and production. My work became a visual exploration of ceramic materials, processes and forms, as well as an examination of how to utilize this information to pursue creative ideas.

Each of Koerth's vessels begins as a wheel-thrown "blank," then is altered and completed by handbuilding.

"Still-life #18 (Spoon Woman Series)," 17 inches in height, earthenware with Cone 010 lithium glaze.

During my undergraduate years, my personal explorations encompassed many processes (low-fire, high-fire, wood, salt and raku kilns) and a variety of materials, but one theme remained relatively constant: my involvement with the vessel. One-of-a-kind covered jars examined as individual forms within a short series, along with bowls, platters and teapots, made up the majority of my work. Pots by Ken Ferguson and Val Cushing offered a standard of achieving strength of statement through form and surface to which to aspire; however, the most important visual and aesthetic influence at this time was Japanese ceramics, especially Shino, Iga and Bizen ware. The gestural quality of line and form, the expressive use of material and process, and the allowance, in fact, insistence for each vessel to have an individual character became my guidelines for explorations in clay.

When I began my graduate studies at Indiana University, I made a conscious decision to change clay bodies (from stoneware to earthenware) and to explore forms from a sculptural direction. Function had never really been an important issue in my earlier examinations of the vessel. (While my jars were certainly functional, my attention was primarily on form and surface.) I realized that my interest was in the visual possibilities of the vessel, not in the functional absolutes. I wanted to relate to my work from a perspective that focused on the vessel as an object and form in space open to any interpretive examination.

My interest in the visual characteristics of non-Western art also continued during graduate school, but shifted from the subtle, material qualities of the Japanese aesthetic to the vigorous, physical expression of Sub-Saharan African and Oceanic art. Indiana University supports a strong African studies program and the School of Fine Arts houses a wonderful collection of art and objects from non-Western cultures.

Research into the art of these cultures reinforced surface qualities—dry, tactile surfaces achieved with slips or low-fire glazes that did not melt and so maintained their applied textures—that had been developing in my work since my last undergraduate year. I enjoyed the softer visual qualities of these dry surfaces, as well as the ability to view forms unobscured by the camouflage of glare and reflection. The textural quality of these surfaces offered a means for visual interaction with the inner areas of the form without interfering with the spatial activity and definition of the outer lines. My concern with the visual integrity of the outer lines of the form grew as I focused more on translating the sculptural concerns of movement and activity in space to the vessel.

"Walking Ewer (Effigy Series)," 17 inches in height, wheel-thrown and handbuilt earthenware.

The vessels that emerged from this concerted effort to change represented the beginning point for my current work. They were full in form, organic in structure and individual in character. Each was an individual expression, begun with a thrown "blank" that was altered, then extended and completed by handbuilding. This forming technique was in a sense like sketching in space, using the material qualities of clay to explore the idea through the building process, which allows the initial idea to evolve according to the particular characteristics of the individual piece. It is a process that still fits my personal needs, allowing me to examine ideas organically, exploring the possibilities as they emerge.

With the "Ritual Vessel" series, I began to explore space, as well as form and surface. The basic elements of the teapot (body, spout, handle, lid) offered ample opportunity to describe and work with negative shapes within and around the form as well as reach out into the space surrounding the object. Through offset placement and orientation of the projecting elements, I was able to break away from the symmetry common with functional forms and establish a degree of gestural movement. The gestural activity of the form, in combination with the attached headlike cup and saucer, implied the human figure and gave each vessel an individual character.

Each "Ritual Vessel" addressed the vessel structure as a beginning point and historical reference. Each also implied the human figure, an additional beginning point in that the vessel itself is a human object. It was, is and may always have been this direct connection between origin and object that has fascinated me most about the vessel.

In the "Effigy Series," the focus of my current explorations, the structure of the vessel and the figure merge together into a continuous form. The body of the vessel implies the figure, while the gestural elements of the figure relate to the visual arrangement of the vessel. The dry, textural surface gives visual substance to the body of the form, while the structure and gesture reach out into its surrounding space. In character, gesture and form, each vessel is an individual. It functions as a visual statement of spatial organization and expression, while celebrating our historical role as vessel makers and vessel users. ▲

"Ritual Vessel (57) III.94," 15 inches in height, by Howard Koerth.

Recipes

Earthenware Body

Cedar Heights Redart	65 pounds
Hawthorn Bond	25
Fireclay	5
Talc	5
Grog	3–5
Red Iron Oxide	1–3
	104–108 pounds

Add approximately ¼ cup barium carbonate per 100-pound batch to prevent scumming.

HK Dry Lithium Base 1

(Cone 010)

Lithium Carbonate	27.55%
Bentonite	3.06
Tile 6	15.31
Flint (Vansil W-20)	54.08
	100.00%

Color variations are possible with the following additions:

Salmon Pink:

Manganese Dioxide	5.00%

Burnt Salmon Red:

Black Copper Oxide	3.50%
Manganese Dioxide	3.00%

Granite Gray:

Nickel Oxide	5.00%

Oyster Shell White:

Black Iron Oxide	3.00%
Vanadium Stain	1.00%

Limestone Green:

Black Copper Oxide	3.00%
Rutile	4.50%

Woody Brown:

Black Copper Oxide	4.00%
Black Iron Oxide	5.00%
Chrome Oxide	0.50%
Manganese Dioxide	3.00%
Red Iron Oxide	4.00%

Bronze:

Manganese Dioxide	2.00%
Copper Oxide	3.00%
Chrome Oxide	4.00%

HK Dry Lithium Base 2

(Cone 010)

Lithium Carbonate	27.55%
Bentonite	3.06
Georgia Kaolin	15.31
Flint (SilCoSil)	54.08
	100.00%

The following are among the color variations possible with oxide additions:

Charcoal Black:

Black Iron Oxide	5.00%
Black Copper Oxide	4.00%
Chrome Oxide	1.00%
Manganese Dioxide	3.00%
Red Iron Oxide	4.00%

Blue Funk:

Copper Carbonate	3.50%

These dry lithium glazes are applied to earthenware fired to Cone 04. To retain application texture and dry surface quality, fire only to Cone 010. Multiple firings may be necessary for color intensity and depth. Shivering may occur if the glaze firing is prolonged. The recommended firing schedule is to turn up the kiln (electric) one third every two hours, so that the kiln is on high in four hours.

A lithium/copper blue is achieved by applying a wash of commercial yellow glaze (Duncan GL600 Buttercup Yellow), thinned with water, over fired Blue Funk, then firing again to Cone 010.

THE DECORATED SURFACE—ADDING PATTERN AND IMAGERY

Many ceramic artists are drawn not only to the process of molding wet clay but also to the wonderful variety of techniques which exist for creating imagery on the surface of clay objects. Just as some painters work with representational imagery and others with abstract patterns; some apply color freely, or even randomly, while others are very precise; so do ceramists take a variety of approaches to decorating their work. The ceramist can use tape, wax, and pencil to plan the design, and airbrushes, watercolor brushes, toothbrushes, and syringes to execute it. Often the ceramist, working on a durable bisqued surface, has more freedom than a painter to sketch and mask out the design before beginning work, or to wash it off if it isn't going well and even to cover a fired, decorated surface partially or completely with a new layer of decoration and refire it. Decorating a three-dimensional surface is a particularly challenging and exciting field of endeavor. On a curved object there is no such thing as a straight line—which may be a frustration to some, but a delight to others. The potter has a choice between bilateral or radial symmetry, or no symmetry at all, and the choice to decorate inside and outside, or front and back, with either identical, complementary, or disparate design elements.

Of course, fundamentally intertwined with these choices are more basic decisions as to what materials to use, and at what stage to do the decorating. Decorating can be integral to the making process, such as with the use of colored clay. It can be applied to the wet clay, by using slip trailing, slip-inlay, and sgraffito, or it can be carved into the wet clay. Decorating can also take place after the piece has been bisque-fired, with the use of underglazes or matt glazes for precise placement of color, or more glossy glazes where interaction between the colors may be desired. Decoration can continue after glaze firing, with china paints, enamels, lusters, and cold finishes. Many of these processes and materials are most appropriate or easiest to use given the predictability and neutral atmosphere of an electric kiln.

The following chapter showcases this range of approach to the decorated surface. The work featured includes functional and non-functional vessels, tiles and "paintings," and sculpture. While flat work, or "paintings" may exist only to show off the decorated surface, other work celebrates the union between the three-dimensional form and its two-dimensional decoration. For most artists who decorate their work, the goal is not to simply "prettify" the surface but to deepen the dialogue between the object and the viewer, to add the language of pattern to those of form and of color. A well-made and appropriately decorated ceramic object invites one to look at it, touch it, and walk around it or pick it up and turn it over. As illustrated in the following pages, it is possible to create such dynamic work with a range of sensibilities, from the playful to the sublime.

The Allure of Slipware

By Irma Starr

I'm a potter, not a sailor, but I was seduced by a mermaid all the same. My particular sea siren reposes on an English slipware plate high and dry in a specially designed case at the Nelson-Atkins Museum of Art in Kansas City. She became my senior project at the Kansas City Art Institute (K.C.A.I.), inspiring subsequent research on and reproduction of 17th-century slip-decorated earthenware.

That mermaid plate is folk art all over, from the uneven, thrown dish to the relaxed, flowing, slip-trailed lines. The Staffordshire potter who created her, Ralph Toft, is so strongly linked in name to English slipware that it's sometimes referred to as Toft ware.

My research turned up few details of Toft's life. There are 35 dishes (recorded to date) with his name as part of the

ABOVE Slipware potter Irma Starr in her well-organized Kansas City studio.

border decoration-the potter's way to identify himself and, perhaps, attract some new customers. The diameters of his plates vary; some measure 17 inches-the average length of a man's forearm, which, for a potter working at a wheel with wet arms and slippery hands, is a convenient way to measure a soft clay dish in progress.

Toft probably was inspired to draw the mermaid, a popular icon in those seafaring days, after seeing her on a tavern sign or a wood engraving. He took plate decorations from everything around him-popular mythological figures like the mermaid or unicorn, religious symbols and the monarchy. He also repeated designs from commonplace objects, such as tobacco wrappers, snuff boxes, etched glassware and pieces of embroidery, as well as birds, animals and flowers.

Framed by a trailed trellis pattern, the mermaid holds the props of her vanity-a mirror and comb. Alongside her are strands of floating seaweed. On that day I first saw her, I entertained the notion that the little lady was admiring herself in the mirror. Later, when my fascination with slipware had led me to study mythology, I discovered the

Original inspiration: a mermaid dish by Staffordshire slipware potter Ralph Toft, 18-inch diameter, circa 1660-1680.

BELOW Plate for the 1989 NCECA meeting, 32-inch diameter, thrown on a fiberglass mold, trailed, glazed overall.

mermaid's mirror is not just a cosmetic item, but a reminder to us that there are things more important to worry about than sagging jowls and baggy eyes.

I laughed when I saw the next slipware plate at the Nelson-Atkins. Made by another member of the Toft family (Thomas), it depicts William of Orange wearing funny pantaloons and pointy shoes. Simple slip lines create a smile that looks as if it could turn into a laugh. The exquisitely drawn hands are graced with dainty lace cuffs to proclaim his royalty. William was a popular king, and Toft may well have picked up his likeness from a "Daniel's Best Tobacco" or perhaps "Escounts...It's Good Tobacco" wrapper.

Nearby, I saw Adam and Eve on another plate. It was the fatal temptation scene-the notoriously sly snake, with an apple in its mouth, wrapped around a tree; the familiar narrative was surrounded by a fence pattern on the rim. My eye then flashed to slipware figures of a "Hare Jumping over His Plate" and "A Mother Pelican in Her Piety" feeding her fledglings. (The mother is said to represent Christ on the cross; the fledglings, mankind.)

Inspired by these pieces, I decided to authentically duplicate the color of the original slips, the clay body, the glaze, decoration and firing techniques used by those early slipware potters. The Nelson Museum's curator of decorative arts, Ross Taggart, who is now retired, allowed me to spend numerous hours in the museum's private holding area, touching the pots, drawing them, tracking the elaborate designs and detailing the various techniques of trailing, marbling and feathering.

He also directed me to the book, English Slipware Dishes, 1650-1850 by Ronald Cooper. This, along with Taggart's ownThe Burnap Collection of English Pottery, and A Potter's Book by Bernard Leach, was an excellent reference. Leach's book was especially important because he and Shoji Hamada were the first potters to rediscover the 17th-century techniques.

Leach wrote: "There was no one, for example, who still had the secret of feathering but [I] eventually rediscovered most of the old techniques. It took me ten years to master slip trailing." I, too, was determined to rediscover these intricate techniques.

The ceramic instructors at K.C.A.I., Ken Ferguson, Victor Babu and George Timock, helped me work out an earthenware body identical to the original: silica sand and fireclay added to Cedar Heights Redart clay. Matching slips to the originals took longer, but through study and experimentation, I was able to replicate the colors.

The original slipware potters coated the inside of each dish with a layer of white slip to provide a surface of satisfactory color and texture on which to decorate. Next, outlines of the design were done in dark brown slip poured from a spouted jug or can. Where narrow lines or dots were required, the slip was passed through quills of different sizes fitted into an opening in the side of the slip-can; the flow was controlled by covering a hole in the top of the can with the thumb. Broad areas were then filled with orange-brown slip, and the original outline ornamented with dots of white slip.

RIGHT "God Bless the Archie Bray," 32-inch diameter, earthenware, with trailed and feathered slips depicting beehive kilns at the Archie Bray Foundation in Helena, Montana (where it was made in 1989), by Irma Starr.

"Mermaid Jar," 14 inches in height, thrown and slip-trailed earthenware, depicting a twin-tailed mermaid, which was a Greek symbol for opposites, such as creator/destroyer, love/hate, etc., by Irma Starr.

BELOW William of Orange portrait (possibly copied from a tobacco wrapper) plate, 18 inches in diameter, circa 1688-1690, by Thomas Toft.

The final step was the application of a lead glaze. In reproducing this glaze, my primary source was Ronald Cooper's book: According to one story, every Friday at a pottery in Derbyshire, one workman roasted pieces of scrap lead over a brazier to calcine the metal oxide. Meanwhile, another workman swept up the dried clay that had dropped to the floor. One bucket of powdered lead "oxide" added to three buckets of floor sweepings made sufficient glaze for a kiln batch. I experimented with ratios of white lead to Cedar Heights Redart clay and found a close match. The iron from the clay gives the transparent glaze a more or less yellow tone, causing the underlying clay to appear cream colored, buff, warm red or rich dark brown.

With the reproduction of Toft's mermaid plate, I completed my senior project and, coincidentally, launched a career. I began to reproduce slipware for the Nelson-Atkins Museum; it was sold through the museum bookstore for many years.

Part of my fascination with slipware has been the challenge of mastering the various techniques. Reproducing the museum's Burnap collection allowed me to become adept at fine feathering patterns, slip-trailing and marbleizing. There is a spirit in these pots that prods me to further experimentation. I am now extending my reach by carving intricate patterns in hump molds and marbleizing variously colored clay bodies.

In 1972, on a research trip to Stoke-on-Trent in Staffordshire, birthplace of slipware, I met Arnold Mountford, director of the City Museum and Art Gallery. He gave me authentic slipware shards, along with old seashells, which were used for the border pattern. I gave the museum my clay and slip recipes.

Last summer, I again visited stoke-on-Trent and met with Mountford, now retired. He introduced me to Pat Halfpenny, the keeper of ceramics, who gave me a complete set of museum slipware slides.

On the advice of these two slipware experts, I visited Winchcombe Pottery (founded by Michael Cardew) and met Ray Finch, Winchcombe's director and potter-in-residence. From Cardew's notebook, he copied the "nontoxic" lead glazes that Cardew used in his later years. Finch also gave me old turkey feathers that were used at the pottery for feathering trailed slip. I sent a pot with one of these sample glazes to the City Museum, where it remains in the permanent collection for reference.

The mermaid still lures me onward. As the Staffordshire slipware potters did in the 17th century, I continue to tell stories on wheel-thrown plates and jars, signing my work with my name trailed across the bottom-still a good way for potters to advertise and, hopefully, attract new customers. ▲

To produce marbled slip decoration, the clay is coated with slip, then contrasting slip is dropped along the edge and the form tilted to encourage flow.

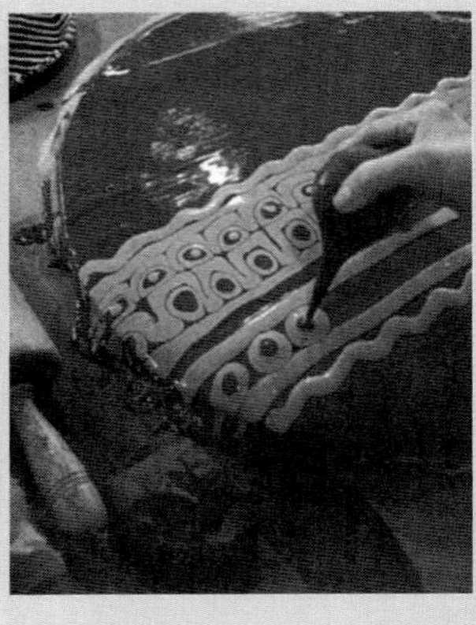

A contemporary version of marbling involves layering lines and dots of light and dark slip; this yields unique hairline separations between the colors.

Feathering, another wet-on-wet process, is accomplished by trailing lines of slip on slip-covered slabs (once the decoration is complete, the slabs are shaped by pressing over hump molds). The tip of an actual feather is used to "comb" (perpendicularly) through the lines of wet slip; Starr uses a raised board as a straightedge and to steady her hand.

Recipes

Editor's note: Traditionally, slipwarepotters used lead (often in its most harmful forms) as a glaze flux because of reliability, color and surface quality. Nevertheless, only those with a thorough knowledge of ceramic toxicology and studio hygiene should undertake experiments with lead glazes.

Slipware Body
(Cone 03)

Barium Carbonate	1 lb.
Talc	18
Cedar Heights Redart Clay	165
Fireclay	37
Silica Sand	37
	258 lbs.

White Slip

Custer Feldspar	15%
Ball Clay	30
Georgia Kaolin	40
Flint	15
	100%

Brown Slip

Cedar Heights Redart Clay	100%

For a dark brown, add 1% manganese dioxide and 3% red iron oxide.

Starr's Lead Glaze
(Cone 06)

White Lead	75%
Cedar Heights Redart Clay	25
	100%

Nonlead Yellow Glaze
(Cone 06-03)

Gerstley Borate	10%
Leadless Frit 3289 (Ferro)	80
Ball Clay	10
	100%
Add: Spanish Iron Oxide	4%

A fired sample of the following lead glaze was determined to be safe according to an accredited laboratory test; the glaze showed a lead release of only 0.31 mg/l (ppm). Potential toxicity to potter or user may vary depending on source of ingredients, studio practices and firing conditions.

Amber Yellow Glaze
(Cone 03)

Gerstley Borate	10%
Lead Bisilicate	70
Ball Clay	20
	100%
Add: Iron Oxide	3%

Apply three coats with a 2-inch multi-head hake brush.

Cardew Clear Glaze
(Cone 03)

Lead Bisilicate	63%
Frit 3134 (Ferro)	13
Kona F-4 Feldspar	10
White Slip (dry)	13
Flint	1
	100%

Apply two coats with a 2-inch multi-head hake brush. For a green glaze, add 1% copper oxide; apply three coats with a hake brush. The green variation tested no higher than 1.48 mg/l (ppm) lead release. The FDA limit for lead in earthenware was 7 mg/l (ppm) at the time of this test.

PHOTOS: JERRY ANTHONY, BOB GOTHARD, STEVE MELTZER

Standing figures, to 7 inches (18 centimeters) in height, handbuilt earthenware, with underglaze and sgraffito decoration.

Washington Ledesma's Parade of Personalities

by Jani Gardner

Entranced with the symbolism of eyes as windows to the soul, Washington Ledesma covers his vessels and sculptures with wide-eyed images.

The vessels and sculptures of Massachusetts artist Washington Ledesma reflect his ebullient spirit and energy. Watching him work is a riveting experience. Whether forming, drawing, carving or painting, it all happens in a quiet fury. With intense concentration, he becomes lost in the moment. Suddenly, pots and figures spring into life: bowls, plates, huge birds on foot-high feet, relaxing rhinoceroses, commanding cats, creatures from other cultures.

Ledesma makes his home on the island of Martha's Vineyard, having had to leave Uruguay and all of his family on short notice during the politically tumultuous 1970s. As a small child, he often visited his cousin's farm, instinctively "playing with mud, mixing it with a little water and straw, making faces and farm-animal shapes. These were moments of great happiness in my childhood; the struggles of my city existence were forgotten."

While in his teens, he studied architecture, and "through drawing made contact with other arts.

Vase, 18 inches (46 centimeters) in height, terra cotta, with underglaze and sgraffito imagery.

At age 24, I enrolled in Uruguay's Bauhaus-inspired School of Fine Arts in Montevideo. I learned art as an exercise of the senses. Painting was food for my eyes, and sculpture fed my touch. I played with fresco murals, jewelry and metals. My first clay objects were primitive human figurines, inspired by Neolithic cave paintings, which generated such an intense feeling of self-discovery that I began an ardent study of early cultures, from Mesopotamia to the Incas. I was starved to know my roots, and ancient art, whether Celtic, Mayan, Aborigine or African, nourished my exploration."

Early on, he was also very influenced by the angular architectural style of Joaquin Torres-Garcia and the Mexican muralists. Soon after, he became entranced with the symbolism of eyes as windows to the soul, and as the soul of the world. At first, these eyes were tiny and tight, then gradually they became larger, wide open. "I was seeing with contemporary eyes, that which was seen by the ancients," Ledesma explains.

Having fled from Uruguay to New York City, he earned a living reproducing "old masters" in a SoHo factory, but soon gravitated back to clay. Working in ceramics gives him a three-dimensional surface for his intensely personal drawings of animals and people.

He approaches each piece individually. "Before I begin to pot, I meditate on all the great potters and painters who have worked before me. I clear my mind until it is blank so their spirits can freely enter. At this moment of openness, the ceramic form begins to grow, without preconceptions or structured guides. The world disappears, except for my two hands holding the clay. When it is over, I am amazed at the shapes that have appeared—like the child in Uruguay who has discovered the joy of playing with mud."

Soon the form is complete, and in the blink of an eye, he cleans and smooths the entire surface with a small sponge and water. He prefers working with white clay for its firing versatility—raku, pit fire or gas kiln—but he is more comfortable with terra cotta.

Large fish sculpture, 26 inches (66 centimeters) in length, terra cotta, with brushed underglaze and sgraffito decoration, once fired to Cone 02.

He then holds the piece in one arm, like a baby, while generously applying color, daubs over daubs, until it covers the clay—and his apron. "At first, I would stand back and look at what I was creating with color. Now I never look down. Color falls where it wants. I am being rather than doing."

After this "action painting," he uses a burnishing technique he learned from Jose Collel, a Spanish artist living in Uruguay. This technique uses four different oils to fix the underglaze to the surface.

Then the pencil drawing begins. Perhaps Ledesma will draw a cat with all-seeing binocular-eyes, or a chicken shape with a face similar to a cat, or elephant angels flying with the birds, which make everyone smile. He might do three-breasted goddesses, to suggest abundance or his mantra about the integrity of life, freedom and truth, or to be, to do and to have.

A parade of pencil-drawn personalities soon marches across the form. "My work is about basic and ancient truths. It is not about the specialization of imitation or the beauty of glazes. The images on the clay grow without doubts, spontaneously, vitally, with plenty of life and color. That freedom is a reflection of the richness of information accumulated in the unconscious mind."

Next comes the carving, or sgraffito, whereby he makes the free-form drawings permanent by carving through the layers of daubed underglaze colors. This is followed by the painting of details: eyes, flowers, hair, etc.

The piece is then ready for single firing to Cone 02 in an electric kiln—three hours on low, three on medium and three on high. When it is cool, he applies a wax finish for protection.

Marketing his work (through East Coast craft fairs and galleries, and from his website www.washingtonledesma.com) requires a good deal of time and energy that Ledesma would, by far, prefer to dedicate to its creation. But then, the real satisfaction of an artist's life is in engaging fellow human beings.

"The only difference between my barnyard figures and my art today," he remarks, "is that now I can relive that magical moment each time someone looks at my creations and experiences a feeling of peace and love. It is as if a cycle that began long ago, before I existed, is continuing: a spiritual gift has passed through my hands and can be received by another. I am utterly thankful to be the instrument of such a gift." ▲

"Standing Figure Female Goddess 'Flower,'" 44 inches (112 centimeters) in height, terra cotta with underglazes, by Washington Ledesma, Vineyard Haven, Massachusetts.

Marty Ray

by Marla Ziegler

"Mama's Kitchen," 19 inches (48 centimeters) in height, wheel-thrown stoneware, with incising, underglazes, glazes and slip, fired to Cone 6.

"Mama Was a Poet," 17½ inches (44 centimeters) in height, wheel-thrown stoneware, with incising, underglazes, glazes and slip, fired to Cone 6.

Seeing an exhibition of Texas artist Marty Ray's illustrated vessels is like turning the pages of a very colorful autobiography, brimming with life events, personal stories and private thoughts. One can't help but taste the home cooking in "Mama's Kitchen," share the excitement of "Installation Day" or hear the rhythms of recitation in "Mama Was a Poet." Ray is quick to point out that "images of life and art have been the primary focus of my work for more than 25 years; and the human figure, within a real or imagined environment, continues to be a favorite subject."

Family, friends and fellow artists appear in Ray's visual stories; her mother is a recurring character. "She was my first art teacher," says Ray. "She taught me how to paint at our dining-room table, and I learned to draw human figures by creating original paper dolls from the Sears catalog. Mama allowed me to paint giant murals on our living-room walls and organize art contests for neighborhood friends. I think of her often when I'm looking for the themes in my work."

Ray divides the making of her work into three distinct phases: throwing/forming the vessel, drawing/creating the narratives, and painting/coloring the images. Using either a brown or white Cone 6 stoneware, she throws the basic form, sometimes joining several together to make vaguely figurative/gestural shapes. Decoration or subject matter is unimportant at this stage. "I don't have the drawings planned for each particular piece," says Ray. "That comes after the form is made."

Her focus on painting, drawing and printmaking in her undergraduate studies

"On the Way to Somewhere," 13 inches (33 centimeters) in height, wheel-thrown white stoneware, with black slip, incised, glazed and fired to Cone 6.

"Free Time," 15½ inches (39 centimeters) in height, wheel-thrown white stoneware, with black slip, incised, glazed and fired to Cone 6.

PHOTOS: LAURIE LAWRENCE, PAT ROBERTS

explains why she uses the verbs "draw" and "paint" to describe the next phase of her work. It was while teaching art in Dallas public schools that she first took a clay class, was hooked, and subsequently continued on to complete an M.F.A. in ceramics and art education at Southern Methodist University. She currently teaches ceramics, sculpture and three-dimensional design at Northlake College in Irving, Texas.

When the clay is leather hard, she lightly draws a design onto the surface so that she can make changes as the story unfolds. "I work from sketches and previous pieces for inspiration. I never have the drawing completely planned. It grows as I go," explains Ray. "Once I'm happy with the total design, I carve the lines deeper, using a metal carving tool with a V-shaped point."

Because of color changes during firing, Ray finds the painting phase (done on the completely dry greenware) the most challenging. "I use a variety of commercial underglazes and color glazes made from a Cone 6 clear glaze that I tint with stains. The underglazes give me a flat surface; the tinted glazes give a glossy surface.

"Working from test tiles, I play with color combinations like I play with pot forms, rarely using the underglazes straight out of the bottle. Underglazes are mixed together to achieve subtle color variations and a variety of light and dark values. I use underglazes as I used to use tempera paint; they feel the same to me."

Ray leaves some areas unpainted, allowing the exposed brown clay to become part of the color scheme. She also may scratch through the underglazes to add details and texture.

After a bisque firing to Cone 04 in an electric kiln, the entire vessel is coated with Blackbird Slip (mixed with a little clear glaze). When the surface is wiped with a damp sponge, slip remains in the most deeply carved grooves, producing dark lines around the figures and objects. On some vessels, these lines serve as "register bars" that not only accent individual scenes but also add complexity to the narrative. Finally, the vessel is glazed on the inside (if needed) and fired to Cone 6 in an electric kiln.

For an exhibition at the Edith Baker Gallery in Dallas, Ray produced monochromatic work, covering the vessels with Dark Slip before incising or carving away lines to reveal the white Cone 6 stoneware beneath. She adapted this technique from her earlier work with woodcut prints. "I make my own slip because I can control the variations of color—sometimes black, sometimes dark brown," Ray explains. "When applied, the slip is about the thickness of milk, and I paint on three or four coats to be sure the color is even. When the slip and clay are leather hard, but not dry, wire loop tools of varied widths are used to carve away the black and reveal the white. I usually start with a few thin outlines and work all over the pot at one time. Paper cutouts of figures are used to repeat shapes, varying their positions as I move around the vessel.

"After a bisque firing to Cone 06, the pot is glazed on the inside with a Cone 6 blue or brown glaze, and the carved outer surface is covered with a very thin coat of Cone 6 clear glaze. I have experimented and know that I need a 'water-thin' coating—just enough to give the surface a slight sheen and to bring out the strong contrast of the black slip against the white clay body. The work is completed with a final firing to Cone 6 in an electric kiln."

Ray finds that the resulting black, white and grays make the new work more abstract. This inclination toward abstraction is also seen in the increasingly stylized figures, making her storytelling more universal than private. Even the titles, "On the Way to Somewhere" and "Free Time," invite you into a broader, collective experience. Whether speaking in a personal voice as she does in her polychromatic vessels or one more general, Marty Ray's forms and images entice us to step inside her world and listen to a well-told tale. ▲

"Installation Day," 19 inches (48 centimeters) in height, wheel-thrown stoneware, with incising, underglazes, glazes and slip, fired to Cone 6, by Marty Ray, Dallas, Texas.

recipes

Dark Slip
(Cone 6)

Potash Feldspar	5%
Ball Clay	75
Kaolin	10
Flint	10
	100%
Add: Black Mason Stain	3%
Cobalt Oxide	2%
Red Iron Oxide	15%

Use on leather-hard clay. Vary the above colorants to yield dark blue, dark brown, etc. Add small amount of frit (such as Pemco P-54) if needed for fit.

Clear Glaze
(Cone 6)

Dolomite	6.60%
Gerstley Borate	21.09
Whiting	2.24
Soda Feldspar	48.77
Flint	21.30
	100.00%
Add: Bentonite	0.96%

Mix to a water-thin consistency; brush over carved, Dark Slip areas.

Never Be Afraid to Play

by Elaine Alt

"Purple Knight," 18 inches in height, wheel-thrown and handbuilt earthenware, with tape-resisted glaze patterns, multifired.

Above all, I want my decorated vessels to convey the tension between whimsy and seriousness. This happens best when I let myself have a little fun with my desire to make things "just right." I've never had an easy time with the line between reality/fantasy, seriousness/ whimsy or loose/tight. I used to believe that one had to choose—loose or tight, never both. After a number of false starts, I've come to think that my pots can be about the struggle to do both. Why not have a piece that is both serious and whimsical?

In my work, the "serious" part is in the balanced forms, geometric designs and craftsmanship. The "whimsy" is in the putting together of things that "don't go": bright gold on matt glazes; hard-edge geometry with squiggles and wiggles; vivid, light-hearted designs on elegant forms. But this is only the most recent work, and it took me a long time to get here. There were many dragons to slay along the way.

I grew up in an ethnic New Jersey home filled with gold cherubs, rainbow prism lamps sitting on pink marble columns, and scenes of Italy painted on the walls. I guess this explains the whimsy. I was also a fairly intense science student and nearly finished a Ph.D. in geology before bailing out. It wasn't until I was in my twenties that I discovered art. I was living in genteel middle-class England, teaching bored English schoolboys how to blow up a science lab. (It got their attention and in retrospect wasn't bad training for future kilns.) Fearing for my sanity, a friend carted me off against my will to evening classes. I didn't know what ceramics was. I had bought my mugs at discount stores. So, along with an assortment of village moms, grannies and kids, I was introduced to ceramics by a very charming, very eccentric English sculptor named John Doubleday. Once I put my hands in clay, like so many of us, I was hooked.

That was the 1970s, when Bernard Leach was revered unconditionally. I enrolled in a ceramics degree course at the Chelmer Institute, and *A Potter's Book* was our bible. We really were taught that there was only one way to pull a handle and one set of propor back in evening classes in England when

tions for a jug. My personal heroes were Lucie Rie and Hans Coper, whose beautiful, sophisticated, elegant works were more than I, an art novice, could hope to emulate. So, I contented myself with trying to make good wheel-thrown functional stoneware the "right" way. Try as hard as I might, though, I could never really master the English ideal of subtle, sensitive work. I had a wonderful teacher, Joanna Constantinidis, who gave me a great piece of advice: just listen to the voice inside yourself and the work will follow. Unfortunately, she couldn't tell me what the voice was saying and I really hadn't a clue that it was those childhood cherubs and prisms trying to get out. I came to value the hallmarks of English ceramics—serious, careful craftsmanship and sensitivity—and still do today. It just took a long time to figure out how to make them work for me.

A few years later, I was back in the states looking at Peter Voulkos' work for the first time. I was shocked and horrified. I could feel the energy, but it seemed so undisciplined. Where was the understated sensitivity of Rie and Coper I'd come to love? Youthful arrogance prevailed and I became determined to show Americans what they were missing. In England, I'd been too intimidated by the "European masters" to tackle porcelain. Back in the states, I dared to place one porcelain bowl on my outdoor craft fair display of functional stoneware. The lone porcelain piece won a prize. Shallow as it sounds, I was thrilled and took it as an omen, went home, gave away all my stoneware and began a decade of producing porcelain pots.

After two years of making wheel-thrown, carved porcelain vessels, I was thrilled when a small New York City gallery (Incorporated, sadly missed since the owners died) offered me a show. The recognition felt great, but the night of the opening, I looked at all those serious pots sitting on pedestals and for the first but definitely not the last time thought the work looked dead. That was quite a shock after all the effort I'd put into the show. A friend, Lucia Jahsmann, gave me my second piece of really good advice: never be afraid to play with the clay.

So, I went back to my studio and thought, I've always been afraid of handbuilding and never made small pots. Maybe I should play around with handbuilding small delicate pots. It was great—just like the thrill I felt being

Black crepe tape is flexible, allowing it to conform to curves.

PHOTOS: DAVID OXTON

"Spider's Web," 16 inches in height, earthenware, with brushed and trailed commercial glazes, dotted with gold luster.

I didn't know what I was doing, and for a while it was lots of fun.

I lost quite a few galleries by changing my work, but it was a small price to pay for the renewed energy of trying something different. I experimented with large, bold and dramatic as well as the subtle and sensitive. Again, I sent out slides and again the pieces grew into something I thought was serious enough to put on a pedestal. Looking back, I see now that, again, there wasn't enough of me in these pots. I had to let go of the idea that my new work had to be serious.

When things aren't right, art has a way of hitting you over the head with it. I got that old dead feeling back. I knew it was time for a change, but I couldn't figure out what. What were those voices saying? What was I supposed to play with? I didn't know, so I learned to ski, hung out at shopping malls and ate too many lunches with friends. If a writer gets writer's block, after 15 years of making pots nonstop, I got potter's block.

For about a year, I just stopped working; galleries lost interest, connections fell by the wayside. Then, one morning, for no reason in particular and without much fanfare, I walked out to my studio, loaded all my porcelain clay and glazes into the back of my van and drove to the town dump. Cleansing—yes. Scary—only a little. Energizing—oh yes. Then I phoned a supplier and ordered some low-fired earthenware and commercial glazes.

In the eight years since, I've learned a great deal visually by looking at the way quilters and other folk artists from Haiti to Africa integrate complex patterns and colors into a whole. I'm inspired by the way Venetian glassmakers blend elegance, delicacy and humor in a single piece. I've also learned a lot on a personal level. I hardly ever ignore the voices of the cherubs and prisms. I know I need to make pots for myself

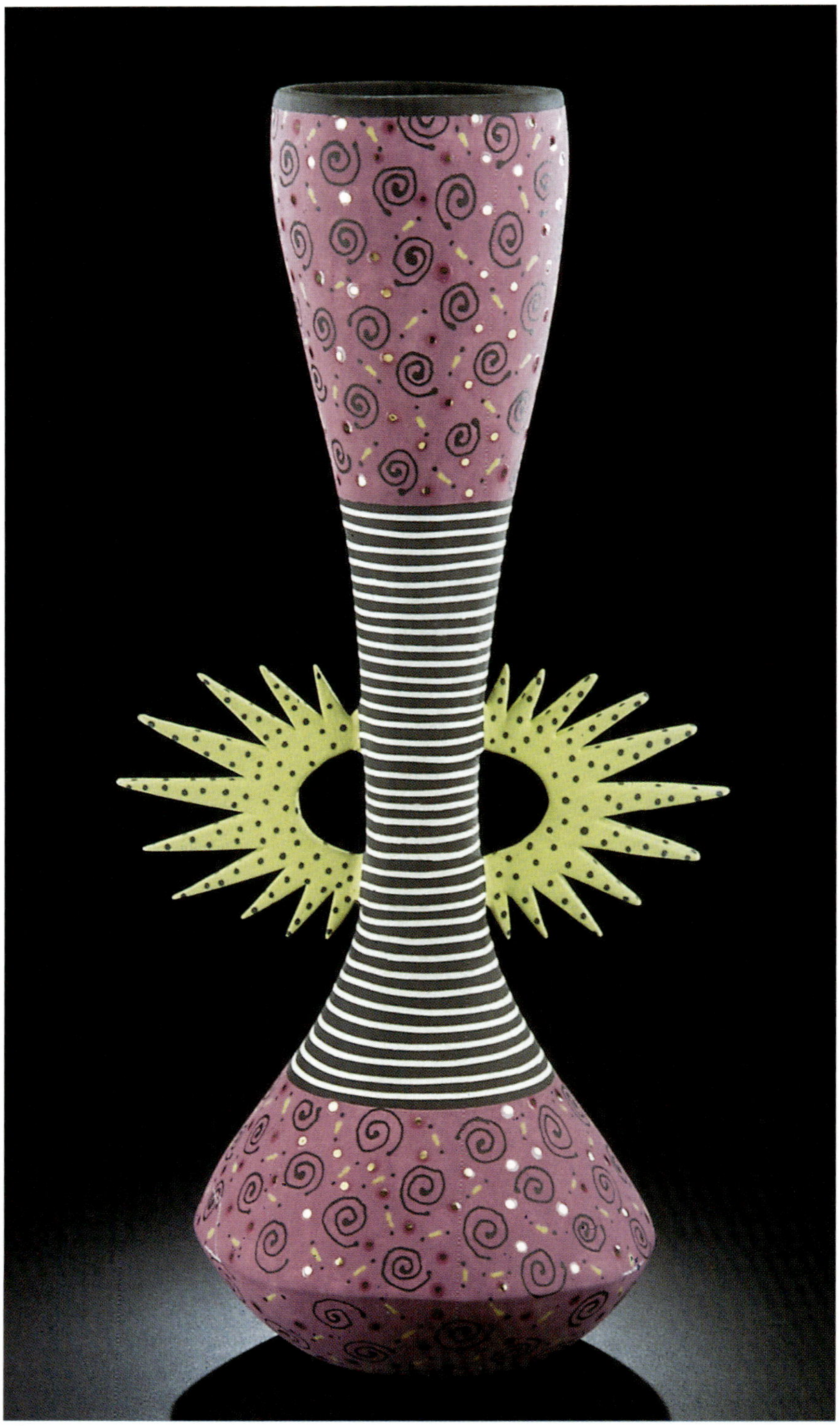

"Galaxy," 17 inches in height, wheel-thrown and handbuilt earthenware vase, with commercial glazes and luster, multifired.

The taping/glazing is repeated as needed to develop complex patterns.

and not for galleries. I've even started a functional line as well, but hey, that's another story about another line it's good to straddle.

My pots are made out of earthenware in a variety of ways. For example, the teapots are both wheel thrown and handbuilt, parts of some bowls are wheel thrown and parts are molded. When dry, they are bisqued to a relatively low temperature (Cone 06), because I find the surface remains more absorbent for the glazing. And then the decorating begins.

In order to achieve the hard-edge lines in the geometric designs, I mask with Charettes black crepe tape. I like crepe tape because it's flexible and I can make it bend into curves. After masking or outlining in tape all the areas to be glazed with one color, I either brush or sponge on two or three coats of underglaze. When the underglaze is dry, I cover it with wax, using the hard edge of the tape to define the wax. Next, I remove and reapply tape where needed to create another color pattern. I repeat this process until eventually all the colors of the geometric pattern are complete and waxed.

The dots, dashes and squiggles are achieved by slip trailing. In order to get rid of the wax and achieve a true bisque, the pieces are fired a second time to Cone 01. For the functional, food-safe pieces, I apply two or three coats of commercial gloss glaze, and fire a third time to Cone 03.

The perfume bottles, teapots and some decorative bowls have a matt finish with gold luster dots. Instead of coating them with a clear glaze, I place a dot of glossy glaze wherever I will want a gold dot. Then, after the piece is fired to Cone 01, I place a dot of gold luster on the shiny spot and fire to Cone 018.

I know the best work happens when I can take the carefully controlled side of myself and mix it up with the whimsical, irreverent side. And when those two come together really well in a piece, it's a great feeling. I also figure that I've been on this most amazing journey with clay and there's no reason to think that this is the last stop. ▲

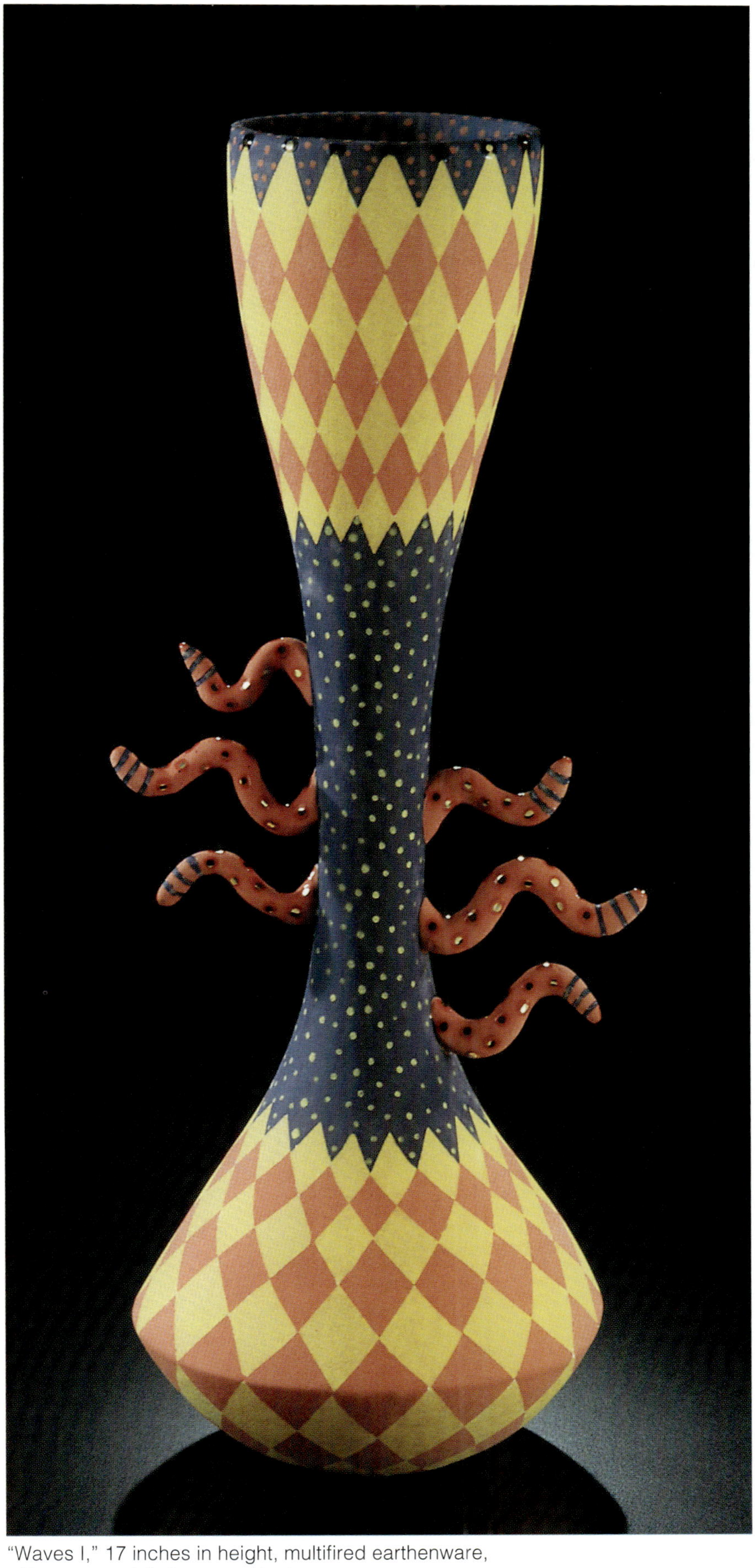

"Waves I," 17 inches in height, multifired earthenware, with commercial glazes and luster.

Perfume bottles, to 13 inches in height, wheel-thrown and handbuilt earthenware, with low-fire glazes and luster, by Elaine Alt, Marblehead, Massachusetts.

Living Tile

by Candy Resnick

When my partner, Ro Mead, and I glance out our studio windows at the breathtaking landscape, we know we have chosen the proper setting in which to do our art. Our studio is located high in the Colorado Rockies at the base of 13,000-foot-high Mount Sopris in a pristine, isolated setting. We are only 25 miles from Aspen, but in a remote area, and don't often have drop-in traffic. We do find that being so far out of the mainstream makes it more challenging to market our tilework, but we gain so much inspiration and nourishment from our environment that the inconvenience is unquestionably worth the sacrifice. Anyone can see by flipping through our portfolio that "colorful Colorado" is not just an empty phrase to us.

PHOTOS: JAMES COHN, GLENN RAND

Candy Resnick and Ro Mead in their studio located at the base of Mount Sopris in the Colorado Rockies.

We each began our individual clay careers more than 20 years ago. Ro moved to this community from Denver in the early 1970s and began teaching pottery at Colorado Mountain College. One of her first projects in the valley was to design a kiln made from handmade bricks. She built the kiln with the help of her students and her potter husband Henry Mead. Her work during this period consisted of a production line of reduction-fired stoneware and porcelain. Meanwhile, I was running around the great Northwest and Alaska, learning the rudiments of my trade by a process of much trial and quite a bit of error, building wood kilns and teaching the Athapascan Indians how to make ashtrays (their choice).

I met Ro in 1979, shortly after returning to Colorado. At that time, we each had an adequate, hands-on understanding of action and reaction as it relates to clay—a solid, if somewhat pedestrian, foundation. Suddenly, however, there were two of us, and evidently there was more that we wanted to do than just push the mud around. We both had been bitten by the color bug.

We have a favorite quote: "The whole is equal to more than the sum of its parts." (Thank you, Mr. Fuller, for putting it so succinctly.) It was the motto we lived by even in those early years when we were still working independently, but sharing our experiments and discoveries. While Ro was fine tuning the technique of appliquéing thin layers of colored clay onto the surface of her vessels, I was utilizing hand-cut vinyl stencils and airbrushed slips to produce sophisticated imagery on my platters.

We were already devoting a lot of energy to perfecting our surface designs. But, eight years ago, when I began building my new home, I incorporated some tile I had made into the bathroom and the work took a quantum leap forward. As I worked my way from room to room, I started breaking out of traditional borders. The concept of "living tile," as we now dub our work, began to unfold. Instead of staying on the wall where it "belonged," the tile began flowing over benches and onto hearths and floors. It had a life of its own; the work had become architecturally integrated and significant instead of merely a utilitarian surface decoration.

Most people have a very narrow view of what and where tile is supposed to be. Because of this bias, we now refer to ourselves as architectural ceramics artists rather than tile makers. This, at least, stops them in their tracks and makes them wonder, "What does that mean?" We are still learning how to deal with this thorny marketing problem of awakening people to the virtually limitless potential of our medium.

Our business partnership began when I received a commission to do a 50-square-foot mural. I persuaded Ro to help me produce it, against the advice of mutual friends who moaned, "You'll ruin your friendship by working together." In fact, the opposite happened. Seven years later, our bond is stronger than ever and the work has grown exponentially.

That first mural was a milestone. Our client had vision. She was clear about her dream and had the unique gift of being able to express what she wanted. Two of the techniques we incorporated into that first piece, at her request, were texture and bas-relief. I

Indian maiden mural installed in Resnick's home; the tile was installed first, then the wall was plastered flush to the tile surface.

A kitchen installation in Hawaii includes three-dimensional underwater flora and fauna, and a tile countertop with induction burners designed to look like coral.

can still remember my husband's horror when he came into the studio one day and we were banging away on an already decorated slab of wet clay trying to give it the texture of rocks. I must admit, it did take guts. For all we knew, we would end up with a bunch of cracked tiles. But from that day to this, we have never been able to resist a challenge. All a client has to do is suggest that we do something that we've never tried (as if we didn't have enough ideas of our own), and we're off and running.

Being raised in *The Little Engine that Could* generation has been a distinct advantage to us in our line of work. I personally must have had that story read to me at least a hundred times when I was a child, so now I always "think I can," no matter how many times the clay says, "Are you kidding? You can't get away with that."

Paulus Berensohn, in his book *Finding One's Way with Clay*, talks a lot about the importance of asking, "What if?" One of the greatest contributions Ro brings to the partnership is her constant "whatifing."

Since neither of us has had a lot of formal training in clay, we have been spared the inhibitions that a more extensive education can sometimes inflict. Let me cite an example to illustrate how too much deference to the "experts" can stifle creativity. Recently, we called a well-known company to ask them why their crimson stain suddenly wasn't giving us the color we needed when it was mixed with our porcelain slip. They asked us for our recipe, and we gave it to them. They replied that they were amazed it had worked in the first place! Fortunately, though, we listened to their analysis of the problem; we weren't willing to accept defeat and we now have the color working again. But "what if" we hadn't been willing to go the extra mile and experiment until we found the answer? Adios, crimson. We follow our intuition and the word impossible isn't part of our vocabulary.

Our particular process is a hybridization of the techniques we were using in our individual work, combined with

A mural installed in Snowmass, Colorado, features 100 square feet of wall tile, plus custom-made sconces that carry the theme of palm fronds into the third dimension.

many new additions. Because of our background, we continue to work on wet clay, in spite of the headaches it causes us. An early distaste for working with glazes on bisqueware was the original impetus for creating our system of decoration with colored slips and clays. We enjoy the immediacy and fluidity of working wet on wet, and we avoid the awkward discontinuity of "retro" decorating. In addition, we also have the opportunity to further enhance our surfaces by sculpting and texturing.

Working large scale requires extensive planning. Our initial presentation to a client includes concept sketches, color samples and bid sheets. After obtaining approval, we enlarge the drawings to a 110% life-size cartoon. The extra 10% allows for grout joints in addition to normal shrinkage. This enlargement becomes our construction road map.

During the next phase, we imagine the finished mural and how to construct it. For example, to suggest rocks, we inlay thin pieces of clay sliced from marbleized clay blocks, then add texture. To portray aspen trees and flower veins, we utilize a wax-paper-resist technique. The surface characteristic of sea sponges and moss is provided by a scouring brush, and precise imagery is accomplished with a series of stencil overlays in a process similar to silkscreening. Each commission is unique and we invariably find ourselves inventing several new techniques for each one.

After we have finished conceptualizing, we mix the colored slips and clays, cut the stencils and roll the slabs we will need for the project. The maximum size we use is 22×50 inches, the largest we can make on our slab roller. After rolling the slabs, we transfer them to Dow SM insulation boards, which provide a lightweight, rigid, nonwarping means of moving the panels around the studio. The slabs are then arranged on our movable worktables according to the configuration of the mural, and we are finally ready to have some fun. We often work on one piece for several weeks, and have discovered that we can keep it wet or even remoisten it by covering it first with a layer of cotton bed sheets, then a layer of wet towels and finally a layer of plastic.

After all the decorating has been done, the slabs are allowed to dry to leather-hard consistency, at which point they are cut into tiles. Each panel is then covered with a sheet of plasterboard to foster slow, even drying, which inhibits warping. When the tiles are bone dry, they are numbered, bisqued, clear glazed and high fired. Firings are all done in an electric kiln to promote brilliant uniform colors.

Our clay body started out as porcelain and has metamorphosed over the years into a high-fire (Cone 8) white stoneware. Though it has lost translucency, it has retained the pure whiteness that is crucial to bright color development in slips and inlaid clay. It has also become an extremely high-performance clay, enabling us to make large free-form pieces, which add three dimensionality to our murals.

A couple of years ago, we did some burner tiles for an induction stove top. When we added some grog and pyrophyllite to the clay for heat expansion, we noticed a bonus we hadn't anticipated—less cracking. It therefore seemed logical to leave these two ingredients in our recipe. This is one example of how important it is to pay close attention to details. Everything that happens has significance. We also found that by leaving grog and pyrophyllite in, we had decreased plasticity; hence the addition of Vee Gum T and vinegar.

Our most recent experiment involves the use of paper pulp. We had read Rosette Gault's article about paperclay in CM [June/July/August 1992] and decided to try it ourselves. We simply added pulp to our regular clay as she suggested. We just got a fired piece of clay out of the kiln that measures a miraculously large 30×40 inches—a testament to the strength of our base clay, as well as the benefits of adding paper.

Our color palette is constantly evolving. The majority of our stains are commercial. Ratios of stain to clay range all the way from 1% to 75%, but generally when testing new stains we find that 12% is a good starting point. We often mix two or three stains together to achieve a precise hue. Recently, we began experimenting with cadmium reds. These are extremely expensive but worth every penny. A good standard for these stains seems to be 30% mixed with the powdered form of our white base clay. We also use the chrome-tin reds and pinks. These can be elusive, but usually a concentration of 75% stain with an addition of 25% whiting under a clear zinc-free glaze will produce rich crimsons, roses and purples.

Ro and I have the rare privilege of being able to work at something that gives us extreme joy. Our philosophy is fairly simple—we create what we love, what excites us and what we think is beautiful. In recent years, there has been an almost ubiquitous suspicion in the art community that beauty might be unfashionable or without depth. It has been riskier to produce a beautiful piece than one that might be garish, ugly and impossible to understand without a lengthy philosophical artist's statement. But a shift toward the revalidation of beauty seems to be occurring. We recently attended an international design conference in Aspen and much of the discussion centered around the preponderance of bad design in our world today—a refreshing affirmation of our own viewpoint. The onus is on us as artists not to add to the ugliness in the world (even with the justification that we are protesting), but to celebrate the magnificence surrounding us, and teach respect and reverence for this fragile and tenuous place we call home.

So it all boils down to education. It doesn't matter whether you get it from a university or life experience; you have a responsibility to pass it on. This business of passing it on seems to be a dominant theme for us. We began by learning from each other in the days of sharing our recipes and techniques. We also feel a strong obligation to teach others about our process. When asked for recipes or guidance from fellow ceramists, we have no hesitation about supplying them with any knowledge we may have gained through the years. We delight in giving classes and workshops, and feel that sharing the excitement we feel for our art always benefits us as much as it does our students. At the same time, however, we caution them not to take our experiences as gospel, and encourage them to try things for themselves. In the final analysis, we are always our own best teachers. ▲

Recipes

Rob's Clear Glaze
(Cone 7–10)

Whiting	15%
Frit 3195 (Ferro)	20
Kona Feldspar	40
Edgar Plastic Kaolin	10
Flint	15
	100%
Add: Bentonite	1.5%

Works well with chrome-tin pink and red stains.

7-A Clear Glaze
(Cone 7–10)

Gerstley Borate	9.71%
Whiting	12.62
Frit 3195 (Ferro)	14.56
Kona Feldspar	38.84
Edgar Plastic Kaolin	9.71
Flint	14.56
	100.00%

RM Pyro Clay
(Cone 7–10)

G-200 Feldspar	15 parts
6 Tile Clay	35
Kaopaque	15
SGP Ball Clay	20
XX Sagger Clay	10
Flint	10
Pyrophyllite	5
Molochite (120 mesh)	10
Molochite (30 mesh)	5
Vee Gum T	1
	126 parts

Add some vinegar for plasticity.

Thomas Orr's Ceramic Paintings

by Kate Bonansinga

PHOTOS: PHIL HARRIS

"Green Wall/Black Door," 14 inches in height, multifired earthenware.

The images that Oregon artist Thomas Orr renders on the back of his ceramic paintings are as important as those on the front. They are reserved for the collector or the brave and curious viewer who takes the time and initiative to lift the pieces, weighty as they are, off the wall and turn them over. These slab-built multifired earthenware "canvases" are about 3 inches deep. Many of them look like terra-cotta bricks in size and shape. Others are larger and squarer, referencing windows.

Orr sees this body of painted wall boxes as having a life span of six years. He predicts that in the next three years the boxes will become flat enough to be framed in metal. At that point, he may add more materials and he may pierce the surfaces.

Most of his recent work incorporates geometric shapes, such as squares, ovals and Xs in bright colors: light blue, brilliant orange, yellow and green. Some pieces have images of house forms; others are landscapes. The houses and landscapes tend to be darker in color. All of them can be thought of as a continuation of Orr's interest in home and history, issues that surfaced in his work in the late 1980s. Conceptually, the weathered, unearthed quality of the front plays off the concealed painting on the back, just as the study of history involves digging up and decoding artifacts or unknown information.

Orr has been working with clay since the early 1970s, but he did not begin painting until he was a student at

Claremont Graduate School (1988–1990). There he was surrounded by painters and felt free to experiment with clay as a painting surface. He also painted on cinder blocks and suitcases. Thus began the sculptures for his "House Series." These developed into his 1990 graduate thesis installation, "Going Home."

The houses in "Going Home" and in the current ceramic paintings may refer to shelter and protection, but they may also refer to confinement or a false facade. While color in many of his graduate-school pieces was less than successful because they were "too rushed, the idea of color wasn't resolved," he overcame this problem by layering slip and glaze color in the body of work exhibited at BonaKeane Gallery in Portland, Oregon.

Orr's process of attaining his distinctive palette supports the content of his work: he creates a history in each of his pieces by layering glaze upon glaze, color upon color. He begins with a coat of white slip on bone-dry greenware, which

"Five Windows," 25 inches in length, earthenware with slips and glazes.

Recipes

Clay Body
(Cone 04)

Barium Carbonate	2 lbs.
Talc	8
Wollastonite	8
Ball Clay	25
Cedar Heights Redart	100
Fireclay	25
Grog (60 mesh)	16
Grog (30 mesh)	10
	194 lbs.

White Slip
(Cone 04)

Feldspar	3 parts
Ball Clay	1
Edgar Plastic Kaolin	6
Flint	3
	13 parts

Apply to bone-dry greenware.

White Crawl Glaze
(Cone 06)

Borax	3.68%
Gerstley Borate	44.13
Magnesium Carbonate	29.42
Zircopax	5.12
Kaolin	17.65
	100.00%

Dry Matt Glaze
(Cone 04)

Barium Carbonate	9.33%
Gerstley Borate	18.65
Magnesium Carbonate	18.65
Nepheline Syenite	27.98
Kaolin	16.06
Flint	9.33
	100.00%

For color variations, add commercial stains.

Chartreuse Glaze
(Cone 04)

Gerstley Borate	5%
Lithium Carbonate	80
Flint	15
	100%
Add: Bentonite	2%
Chrome Oxide	3%
Tin Oxide	7%

Lizard Glaze (Black)
(Cone 06–04)

Borax	6.3%
Gerstley Borate	43.8
Lithium Carbonate	9.4
Magnesium Carbonate	25.0
Nepheline Syenite	12.5
Flint	3.0
	100.0%

acts as the equivalent of gesso on wood or canvas. He then bisque fires the pieces in an oxidizing electric kiln.

On the abstractions and houses, Orr applies several colors of glaze, each isolated from the other, and fires to Cone 04, again in an electric kiln. After this first glaze firing, the surface has a glossy, wet quality. He then applies more glazes, perhaps the same ones as the first time, perhaps different ones, and fires a second time to Cone 06.

For a dry, flaky appearance, he uses glazes with a high magnesium carbonate content. In certain places the glossy, initial layer of glaze shows through, which conveys a sense of age and depth. Orr then adds more glazes and fires at least a third time, often more, repeating the layering and firing as many times as necessary to achieve the desired surface texture and color. He sometimes completes a piece with a cold finish of wax, paint or polish.

His approach is somewhat different for the landscapes, which are new to his repertoire. If the abstractions and houses are mature, the landscapes are in their infancy. On these long, narrow, horizontal pieces, he mixes the glazes directly on the surface, rather than applying each color separately in a discrete area.

This technique is reminiscent of a painter blending colors on canvas, an activity that may be the inspiration for the final sentence in Orr's artist's statement: "Sometimes I feel like a painter trapped in a potter's body."

"Three Sons in the House," one of the pieces in Orr's exhibition, synthesizes the concepts of home, landscape and abstraction. It is a wall box that refers to a window that offers access to the world beyond oneself. The background is pale green and is surrounded by a bold, black painted outline. Three orange spheres in the upper left corner float above a yellow elliptical shape balanced sideways on five black legs. To the right is a large, black rectangle within which is a hint of a pointed roof line. Orr intends for the viewer to "understand the work through his or her own unique insight," just as we understand our and others' histories, experiences and homes. ▲

"Three Sons in the House," 18 inches in length, earthenware, with brushed slips and glazes, fired to Cone 04 in an electric kiln, by Thomas Orr, Portland, Oregon.

Sun Chao: A New Way with Crystals

By Tobie Meyer

"Return from Paris #4," approximately 28 inches in length, a crystalline glaze "painting" on ceramic "board," by Chinese ceramist Sun Chao (right).

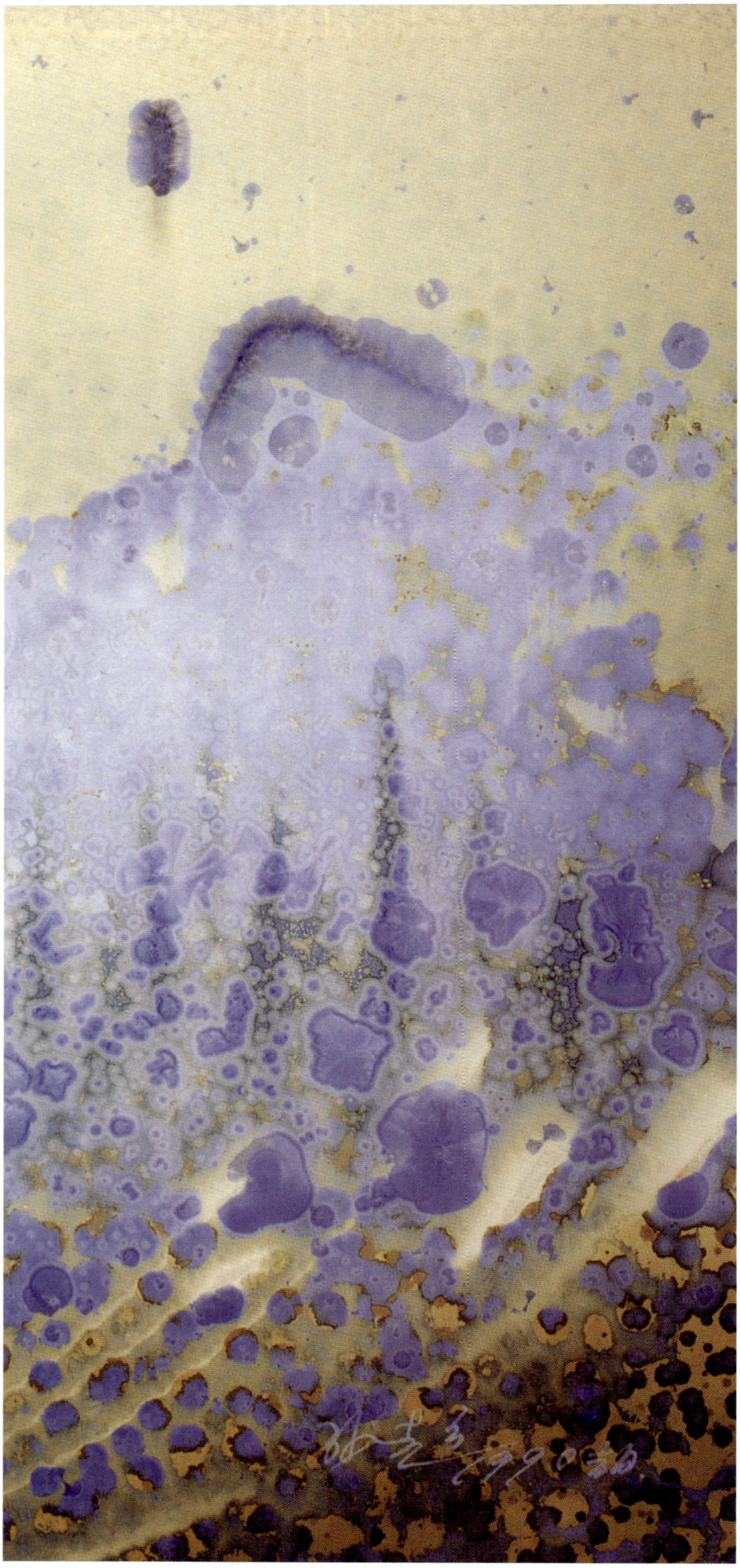

Sun Chao, a Chinese ceramist living in Taipei, has explored the possibilities of aesthetic expression through the medium of crystalline glazes for more than 20 years. He likens the continual evolution of the crystals inside the kiln to the ongoing personal changes that take place over an individual's lifetime, seeing his vases, plates and glaze paintings as visual metaphors for the human experience.

Typically, crystalline glazes are characterized by the beauty of the crystals themselves, but Sun endeavors to bring more to his glazes. By spraying, brushing, pouring, layering and scraping, he manipulates the results. Consequently, he can improve upon the typical crystalline glaze by overcoming the tendency of crystals to cover the entire surface.

Having moved beyond an exclusive focus on crystals, Sun has opened a broader range for ceramic expression, sometimes incorporating the sense of freedom, depth and fluidity found in Oriental ink paintings, and at other times incorporating elements resembling forms found in Western abstract painting.

During the 1970s, Sun played a major role in the establishment of the Conservation Division at the National Palace Museum in Taipei. He has since devoted much time to researching ancient Chinese glazes, including Neolithic; Tang-dynasty (618-906) tri-color; Song-dynasty (960-1279) celadon and temmoku; and Ming-dynasty (1368-1644) copper-red and blue-and-white glazes. This long-term study provided him with a breadth and depth of knowledge that, combined with his artistic abilities, allowed him to pursue his new vision of crystalline glazes. ▲

"Return from Paris #7," approximately 28 inches in length with sprayed, brushed, poured, sprinkled and scraped crystalline glazes, fired to 2370 degrees F, then soaked at 2010 degrees F for five hours.

"Return from Paris #5," approximately 28 inches in length, layered and manipulated crystalline glazes on ceramic board, by Sun Chao, Taipei.

MAJOLICA—FROM THE RENAISSANCE TO THE SPACE AGE

One of the most widely used decorating techniques among electric kiln potters is that of majolica. Developed originally on the island of Majorca in the Mediterranean, in the 15th century, majolica involves decorating with stains or oxides over a tin-based, white glaze, and is characterized by a slightly softened line and grainy color due to the melting of the colorant into the glaze. Originated in an attempt to imitate the whiteness of oriental porcelains, majolica is now popular as a way to create a visually rich surface on oxidation-fired pottery.

Traditionally, majolica is an earthenware technique that uses a red claybody and a lead-based glaze. Most modern practitioners continue to use red clay, but lead-based glazes have fallen out of favor. The development of reliable colorants in the form of commercial stains has made majolica a very accessible technique, though mastery of majolica requires much practice. Painting on a powdery, already-glazed surface can be tricky, and there is no room for error. Individual brush-strokes are evident, so brushwork must be sure and precise. Potters who take up the practice of majolica often spend a lifetime perfecting the technique and exploring its possibilities.

Though the bright colors typical of majolica ware, as well as its development in the warm Mediterranean climate, often lead potters to decorate with depictions of fruit or flowers, there is no rule requiring potters to limit themselves to nature-based imagery. As practiced now, the age-old method of majolica remains the same but the motifs range from the sunny to the somber. However, majolica lends itself to a certain playfulness, both in the decoration and in the forming of the clay, which requires a great deal of sensitivity on the part of the potter. As this chapter illustrates, majolica is overwhelmingly used to decorate functional pottery, though, as with any earthenware technique, this necessitates solving problems of porosity and glaze fit. The articles in this chapter illustrate a few of the many possible ways to take advantage of this versatile technique.

Invisible Gesture

by Paul Rozman

Majolica-glazed coffeepot, 10 inches in height, 1996, by Paul Rozman, Calgary, Alberta, Canada.

Along with the expressive nature of the material and ideas of form and utility, there is an overwhelming presence in the pots I admire most. This presence speaks of human experience, regardless of time and space. I call it invisible gesture because it is not manifested in a physical way but speaks directly to our guts, regardless of cultural barriers.

Contemporary pottery production is not strictly inspired by market demand or by a dominant aesthetic. But this new-found freedom can be a disadvantage in the absence of mainstream criticism.

The challenge for me is to develop criteria that are not necessarily universal but dare to be individual and relevant to my own experiences. My intent is to understand and to find meaning in what I do. This is why I choose to think of criticism as ceramics appreciation.

Functional pottery involves ideas and concepts, including form painting and/or form decoration, but it also deals with an aesthetic of use, a unique opportunity to experience art in the most intimate way. When the user takes hold of a coffeepot's handle, the mundane becomes a sensual engagement, protecting the user against indifference. I introduce subtle hints of the forming

Shallow bowl, 15 inches in diameter, wheel thrown, with majolica glaze, fired to Cone 2 in oxidation, 1995.

process on knobs for the same reason. This interaction between the user and the pot, through sensual awareness, constitutes but one aspect of the potter's art.

Another aspect of functional pottery is its potential to offer process, form and function as an idea of beauty. The universal reaction to a natural splendor is to gaze in awe. The results of ceramic processes are beautiful because of fidelity to natural phenomena. Through study of historical examples, I see variety and cultural differences in this relationship with nature. In my majolica work, I use images of familiar animals (moose, cow, chicken, fish, white owls and frogs) integrated with stylized background patterns of water, trees, leaves or simple color fields. The majolica technique lends itself readily to this kind of approach to surface. It is possible to have a full range of depth from the white glaze to transparent washes of color to opaque saturated colors. Regardless of color saturation, the painted surface melts into the glaze and therefore takes on its character. This makes it possible to have rich surfaces without interfering with the utilitarian purpose of the forms.

The notion that functional pottery is too limiting for expressive needs is a strange one to me. Unfortunately, our culture has adopted the belief that func-

Wheel-thrown platter, 14 inches in diameter, majolica glazed, fired to Cone 2, 1996.

Slip-cast cups, 3 inches in height, with brushed pigments over Rozman Majolica Base Glaze, 1996.

Recipes

Rozman Majolica Base Glaze
(Cone 2–4)

Whiting	5 %
Frit 3124 (Ferro)	40
Kona F-4 Feldspar	15
Nepheline Syenite	15
Edgar Plastic Kaolin	12
Flint	13
	100 %
Add: Superpax	12 %
CMC Gum Solution	10 %

The gum solution is a mixture of 15 grams CMC gum in 1 liter water.

Rozman Stoney Matt Base Glaze
(Cone 2–4)

Whiting	20 %
Frit 3124 (Ferro)	20
Nepheline Syenite	30
Edgar Plastic Kaolin	15
Flint	15
	100 %

For a yellow variation, add 5% rutile; for blue-green, add 0.3% cobalt carbonate and 2% copper carbonate.

Rozman Throwing Body
(Cone 2–4)

Wollastonite	3 %
Nepheline Syenite	14
Ball Clay	25
Hawthorne Fire or Cedar Heights Goldart	25
Cedar Heights Redart	25
Flint	8
	100 %

Rozman White Casting Slip
(Cone 2–4)

Talc	5.00 %
Nepheline Syenite	30.00
Edgar Plastic Kaolin	23.00
Kentucky Ball Clay (OM 4)	22.00
Flint	20.00
	100.00 %
Add: Barium Carbonate	0.03 %
Soda Ash	0.03 %
Sodium Silicate	0.25 %

Start by adding 38% water, then add more if required.

Matt-glazed bowl with iron brushwork, 12 inches in diameter, by Paul Rozman, 1996.

tion is synonymous with convenience. This notion is so prevalent in modern design that often we confuse the two. A Styrofoam cup is perfectly satisfactory when the concern is convenience alone. This solution, though, is so one-dimensional that the only pleasure I get out of using it is when I throw it in the garbage. It is nevertheless an aesthetic of convenience, simple-minded and restrictive.

An important component of the human experience is the opposite pull of intellect and emotion. One of the major attractions I have to functional pottery is that this apparent conflict is welcomed. In fact, I will go further to say, not only are these opposites fused in pottery, but I see this as an integral part common to the pots I admire most.

It is a given that art pottery involves ability, intellect and emotion. Functional pottery, by its nature, requires a level of ability (process and technique) to fulfill concept and subject matter (which are solidly rooted in life experiences). The soul comes from our traditions and firsthand experiences. The degree to which all these human qualities are involved and expressed in an individual manner makes the difference between utensil and art object. ▲

Morgen Hall

"Salt and Pepper Set," 5½ inches in height, earthenware, with slips, tin glaze and stain.

Wheel thrown from red earthen ware, Wales potter Morgen Hall's place settings, serving dishes, tea sets, salt and pepper servers, and storage jars are often detailed with fine rouletted impressions (a design element that comes from British tablewares of the 18th century). Paper-resisted slip allows for crisp-edged patterns, while trailed slip adds surface relief. Over the creamy white tin glaze, she sponges a rutile-based, mustard-yellow glaze stain. The ware is then fired to 1120°C (2048°F) in an electric kiln.

"Coffee Cabaret Set," to approximately 11 inches in height, earthenware with resisted and trailed slips, tin glaze and rutile stain.

PHOTOS: PATRICIA AITHIE, PHIL MARTIN

"Ginger Jar," 15 inches in height, wheel-thrown earthenware with slips, glaze and stain.

"Tea Cabaret Set," to approximately 8 inches in height, earthenware with slips, glaze and stain.

"Serving Platter," 19 inches wide, wheel-thrown and handbuilt earthenware with slips, glaze and stain, by Morgen Hall, Cardiff, Wales.

"Cinnamon, oranges, cream and mustard are used to describe the work, and the food connections with my tableware are essential, as all of the work is intended for everyday use," Hall explains. "I strive to make work that is a pleasure to use, where handles are inviting to pick up and spouts are eager to pour. I have a positive enjoyment in making the work, which I hope translates into a direct enjoyment by others of using the work.

"I have an ongoing interest...of past traditions in ceramic tableware throughout the world," she concludes, "but equally a commitment to trying to make work that is of our time and place." ▲

Steve Davis-Rosenbaum

by Nancy K. Foreman

"Chip and Dip," 12 inches in height, majolica-decorated terra cotta.

Steve Davis-Rosenbaum is a traditional potter with a fine artist's eye for design and color, bringing a complexity and an intricacy to utilitarian objects.

"Much of my pottery originates," says Davis-Rosenbaum, "in the basic human joys of eating and cooking with all their overtones: fireside, nourishment, caring and celebration. Beautiful dishes have become synonymous for me with love of food and its presentation. Even my vases present themselves to me in the context of a table setting. I make my pots based on my personal domes tic needs and lifestyle choices; and the focus of my production is everyday dishes for use in cooking, dining and home decoration."

Over a majolica base glaze applied to the wheel-thrown terra-cotta forms, he applies a variety of stains that react with one

PHOTOS: GEOFFREY CARR

"Triple Jalapeño Dip," 8 inches in height, wheel-thrown terra cotta, with fluxed stains on majolica glaze.

"Covered Jar," 8½ inches in height, with brushed stains on majolica base glaze.

"Open Vase," 7½ inches in height, majolica on terra cotta, by Steve Davis-Rosenbaum, Lexington, Kentucky.

another in the kiln. Most combine a tin/vanadium stain with a chrome black or green stain to encourage chrome/tin "flashing."

Though salt-glazed stoneware has been Davis-Rosenbaum's mainstay for the past ten years, this new flashed majolica reflects an Italian influence (in the majolica-type white glaze) meeting the influences of Japanese and Islamic bold decoration.

Davis-Rosenbaum admits that he is still "struggling with the same issues of form, function, surface, decoration and color" that all evolving artists know so well. Art must, of course, embrace emotion and intellect, but to bring to fruition any given idea or theme in functional pottery, an artist must also understand process and technique. ▲

"Bowl," 4 inches in height, wheel-thrown terra cotta, with brushed stains on majolica base glaze.

Flashed Majolica

by Steve Davis-Rosenbaum

All my pots are thrown on a Leach-style treadle wheel, then altered off the wheel. Each is covered with a white majolica-type base glaze in preparation for painting with colored stains mixed with a flux.

I prefer to use stains that will react with one another, causing flashing effects. Usually, I begin with a tin/vanadium layer that turns pink/orange/red, depending on the flux used (Gerstley borate or Ferro Frit 3124) and add chrome black or green patterns. The resulting chrome/tin flashing adds variation to the surface.

The base glaze is Linda Arbuckle's majolica recipe:

Linda Arbuckle's Majolica Glaze
(Cone 04)

Frit 3124 (Ferro)	65.72%
Kona F-4 Feldspar	17.23
Nepheline Syenite	6.24
Edgar Plastic Kaolin	10.81
	100.00%
Add: Tin Oxide	5.00%
Zircopax	10.00%
Bentonite	2.00%

Over this I apply a variety of commercial stains. The tin/vanadium yellow stain that I use in combination with chrome stains to get "flashing" was previously sold by Leslie Ceramics, but is no longer available. I stocked up a few years ago, and am still using what I have left. Presently, I am in the process of testing a variety of other yellows made from tin/vanadium combinations. Initial tests have been positive with Mason 6440 or 6404.

Some of the other stains I use are Leslie Z-560 for black, Mason 6223 for green, Mason 6605 for crimson and Mason 6254 for blue-green.

For onglaze decoration, these are mixed 1 teaspoon of stain to 3 teaspoons of Gerstley borate, or 40% stain to 60% Ferro Frit 3124. To improve brush flow, I add some Veegum/CMC/water (a 1:1:2 mixture recommended by Walter Ostrom).

Contempo-Baroque Majolica

The Art of Simona Alexandrov

by Louise Melton

PHOTOS: W. MARC BERNSAU, DB PHOTOGRAPHICS

"To Your Male Admirers," 4 inches in height, wheel-thrown whiteware, with majolica glazes and lusters, fired to Cone 04.

Swedish traders called the river where Russia's Peter the Great would construct his Baltic port *neva,* their word for mud. Raised up out of a frozen swamp by slave labor, Italian architects and the autocratic will of Peter and, later, Catherine the Great, Saint Petersburg became the site of some of the finest late baroque architecture in Europe, including the splendid Hermitage Museum.

Two centuries later, the name of the city would be changed to honor the memory of Lenin. But the grimy industrial colorlessness that so characterized other major Russian cities during the Soviet period never really settled down to live in Saint Petersburg.

When ceramics artist Simona Alexandrov was an 11-year-old walking along the streets of Leningrad on her way to art lessons, she was fascinated by the rich gray-green copper domes and dynamic facades that seemed more like sculpture than architecture. Endless sub-Arctic white nights crystallized color into piercing flashes of iridescence off the snow, glazed branches and pinpricks of quartz in the granite blocks that line the River Neva. If Alexandrov walked in one direction away from her family's small apartment, she passed a symphony hall. The energy and expansiveness so characteristic of the baroque period (1600–1750) permeated the child's world and helped define what is beautiful for her. As an adult, Alexandrov would celebrate Saint Petersburg's architecture, its river and sky, its statuary, art and spirit in vivid majolica-glazed cups, vessels, wall pieces and ceramic sculpture.

"White Night Reflections" (page 94), for example, invites the viewer into a deceptively whimsical world where Saint Petersburg's "architecture" becomes a visual pun for the "architecture" of the piece or, as Alexandrov explains, "the beauty of thought through composition." The main body of the cup presents a row of city buildings which, at first, appear as tidy and restrained as the buildings in Amsterdam that greatly influenced Peter's plans. The base of the teacup is banded by the Neva's stone walls. The saucer is the reddish, muddy Neva itself. As one turns the cup, however, one experiences a subtle discomfort. What is going wrong here? The buildings are cunningly reflected in the river and if the cup is not in position, the piece immediately loses its compositional logic. The sculpted figure of the Goddess of the White Night on the lid wears a cobalt blue cape studded with gold luster stars. Unlike the city, this artful "sky" can be freely rotated into any position in spite of the red reflections beneath the figure's cape.

Consciously employing the viewer's personal and historic sensibility in the unity of the piece is fundamentally a baroque technique. Alexandrov's work always has an inviting narrative quality but one that is open rather than specific.

Alexandrov is now a studio artist and teacher in the Boston metropolitan area. "I don't understand why people ask me to interpret the stories on my work," she says. "I do not want you to say, 'I must accept the story the artist intended.' If the work is successful, you experience *your* story there. I have to anticipate that in the composition. I am like a chariot driver who has to control six horses—drawing, painting, architecture, form, fire and technology—all to get to the same place. The viewer is like a seventh horse, one I do not know so well. I cannot control this horse. But all are yoked together, so I sense and work with the seventh horse, too."

Alexandrov's conscious union of multiple disciplines is also a hallmark of the baroque style. It was an early fascination with the integration of architecture, color, sculpture and theme that prompted an insightful art teacher to tell Simona's mother that her child was in the wrong class. "Your daughter paints like a ceramist," he said.

"So the young student who was just going along without a care thinking it would be okay to be a painter, just as carelessly decided to follow this advice," Alexandrov recalls. "He was right. Ceramics suits me. To have so many compositional elements to work with is more complex, but complexity interests me.

"I am lucky," Alexandrov says today. Her family emigrated to the United States in 1988 after ten years of waiting for exit visas. "That was before *perestroika,*" she says. "My parents both lost their jobs just for applying to leave Soviet Union."

The luck Alexandrov refers to with a laugh, however, is the fact that she was accepted at Saint Petersburg's Muchina Museum School for the Applied Arts. "I was 19," Alexandrov says. "If I had understood how competitive and anti-Semitic it was, especially in the ceramics department, I would never have applied. It was good for me that I was stupid. It was good for me that I had a mentor in Viktor Petrov."

Petrov is the painting teacher and ceramist who recognized Alexandrov's talent and fostered it, in spite of her family's lack of influence with the authorities who decided who did and who did not get into prestigious schools.

The art academies in Saint Petersburg and Moscow had, at the time, what Westerners view as a restrictive emphasis on historical tradition in service of the now infamous Socialist Realism. The large state academies called the modernist aesthetic of the early part of the century "bourgeoise formalism" and rejected it. The Muchina, however, actively preserved the ideas associated with the fauvists, cubists, expressionists and constructivists.

Alexandrov's teachers passed on a tradition of innovative explorations into the interaction between structure and emotion. Writing later, Alexandrov would cite expressionist Wassily Kandinsky and constructivist Kazimir Malevich as major influences, along with the classical discipline she learned at the Muchina, in her development as an artist.

"In the United States, students usually start working right out of a bag of clay. Before I went to the Muchina, Viktor Petrov had already taught me to start with composition. Before I touched clay,

"On the Roofs of Baroque," 7 inches in height, with brushed majolica glazes and lusters, fired in oxidation to Cone 04.

I had to produce one simple sketch that combined the drawing, painting and structural elements of how the piece would stand and function. Viktor and, later, the teachers at the Muchina stressed an alternation of analytical and spontaneous stages, but everything had to make sense in the sketch or it wouldn't make sense in the final piece."

This rigorous methodology is what makes all the elements of Alexandrov's complex work flow together in an underlying and reassuring formality, no matter how frivolous the themes may appear to be at first glance. "Jester Vessel" is a good example of Alexandrov's control over a vibrant pastiche of colors, images and forms. Four pieces were thrown to produce an immediately familiar vase structure, then altered to give the body of the piece a jaunty cocked-hip attitude that reinforces the painted jester, *tragediennes* and jugglers.

Theater is a frequent theme of Alexandrov's. The acrobats on these works are reminiscent of the commedia dell'arte figures on 18th-century Sevres faience, of the *istoriato* style developed in Faenza at the beginning of the 16th century, which introduced the human figures as the central focus of a scene, and of Alexandrov's frequent attendance at the theaters near her childhood home where she went, as she recalls, "to sketch, think and try out new ideas."

Simona Alexandrov would discover her own baroque contemporary style by expanding on ceramic techniques she was able to study firsthand at the Muchina. The Muchina was founded at the end of the 19th century as one of the first museum schools to accept women. It was then called the Stieglitz School for Technical Drawing for Both Genders. The allied Muchina Museum housed one of Europe's finest collections of majolica.

The Hermitage took over the smaller museum's collection to protect it during World War II. The "protector" kept the best of the Muchina's majolica collection for itself after the war ended; however, a number of historically significant pieces that were not important for the Hermitage collection were returned when the Muchina reopened in 1944. Students at the Muchina were encouraged to study them in detail.

Simona Alexandrov experimented with reproducing techniques handed

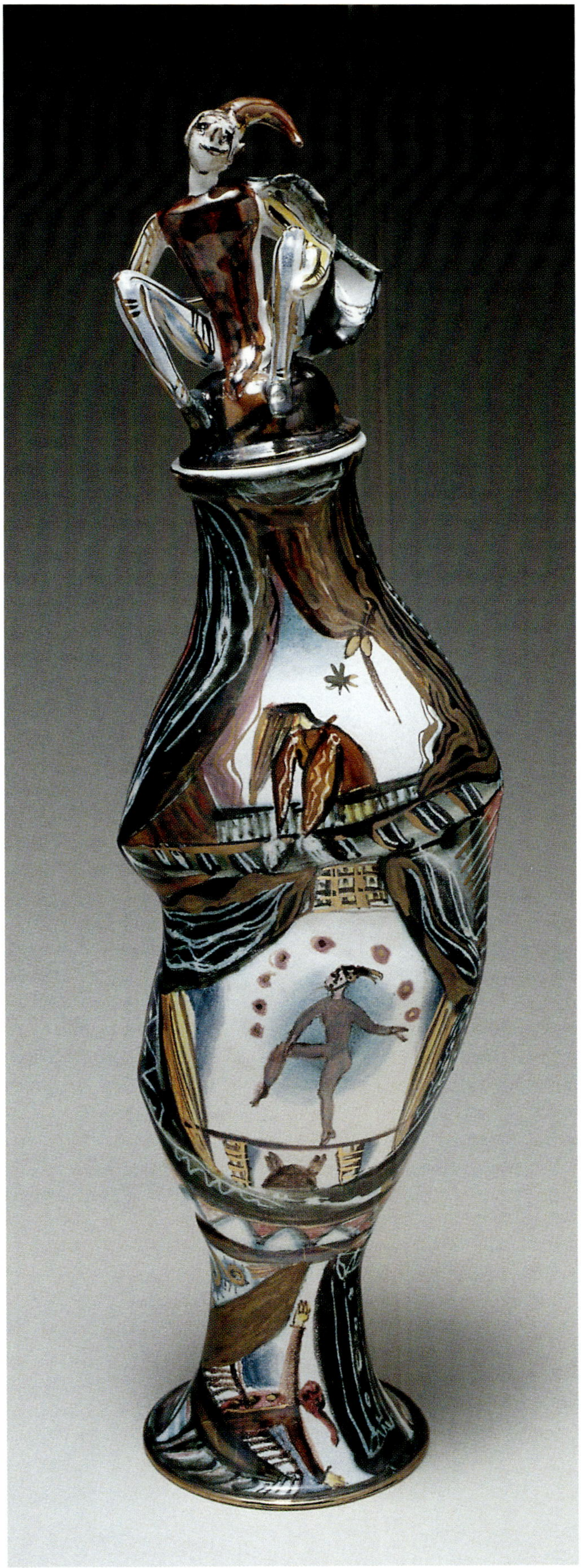

"Jester Vessel," 15 inches in height, wheel-thrown and assembled whiteware, with majolica brushwork.

"White Night Reflections," 10 inches in height, wheel thrown and handbuilt, majolica-glazed whiteware, by Simona Alexandrov, Waltham, Massachusetts.

down from 10th-century Moorish lusterware, 15th- and 16th-century northern Italian faience and 18th-century French majolica, as well as the Russian ceramics workshops centered in the Gzhel region outside Moscow in the 1700s. "Museum pieces usually are untouchable," Alexandrov remarks. "You can look at them and read about them. But the chance to hold a priceless antique majolica pot in your hands and really see how it was done was an incredible experience for me. We copied these ceramic pieces the way painting students copy the old masters—in order to experience the techniques they used and truly understand them."

Soon after reaching the United States in 1988, Alexandrov was awarded a full scholarship to study for an M.F.A. at Cranbrook Academy of the Arts in Michigan. There, she found encouragement for her personal voice from Cranbrook's faculty and visiting artists. "Cranbrook is very much about doing your own work in your own studio," Alexandrov remembers. She credits this supportive environment for a new level of universality in her work.

Currently, her work is created through a series of stages—usually eight or nine but sometimes more, depending upon the complexity of the structural elements and brushwork on the piece. She always uses white clay instead of the traditional terra cotta of Italian majolica.

On the dried white glaze, Alexandrov first applies sulfates and underglazes. "Sulfates can be very speculative in terms of color but really allow the painting—brush strokes, lines and shading—to come through," she says.

Alexandrov uses sulfates in a 10% wash. Because sulfates are highly toxic, she uses them in a well-vented area and never sprays them. Sulfates may be brushed with complete freedom but because they become invisible after they dry on the white glaze, Alexandrov often adds ink or watercolor to the sulfate to help keep track of her painting. Once the underglaze and sulfate painting is complete, Alexandrov sprays the piece with a clear glaze.

The piece then undergoes as many as five or six Cone 04 oxidation firings to make changes, fight pinholes and correct defects. It is sprayed with clear glaze before every firing.

When Alexandrov is satisfied that she has achieved the overall design of the painting she envisioned, she paints over selected areas with metallic lusters. She then fires the lusters to Cone 022–017.

Lusters enliven Alexandrov's painted worlds with scintillating light reminiscent of the way Saint Petersburg's white nights paint the city with brilliance in the summer. "Lusters give you a depth and resonance of color like the Rembrandt effect," she explains.

The advantage of lusters is the ability to edit the majolica paintings at will. Her favorite historical examples of this are commemorative wedding plates. "They were made for rich families in Italy and handed down to the sons and daughters when they married. Rather than have a new plate made at considerable expense, the faces, names and dates of the bride and groom would simply be painted over with lusters to update them for the occasion. Because the pieces were low fired, the colors underneath were not disturbed."

Alexandrov says that when she is designing and painting a work, she is like an architect—controlled and precise. "But," she adds, "the combined effects of the underglazes, sulfates and lusters are a little like the spontaneity of the watercolorist. When the fired piece comes out of the kiln, there can be surprises." ▲

Recipes

Off-White Majolica Body
(Cone 04)

Talc	13 %
Frit 3124 (Ferro)	4
Cedar Heights Goldart	25
Fireclay	40
Kentucky Ball Clay (OM 4)	10
Flint	8
	100 %
Add: Fine Grog	10 %

Majolica Glaze
(Cone 04)

Talc	9.53 %
Frit 3124 (Ferro)	37.14
Frit 3195 (Ferro)	37.14
Kaolin	7.62
Flint	8.57
	100.00 %
Add: Zircopax	9.53 %

Clear Glaze
(Cone 04)

Frit 3124 (Ferro)	80 %
Nepheline Syenite	10
Edgar Plastic Kaolin	10
	100 %
Add: Bentonite	3 %

Majolica-decorated earthenware tiles, each 6 inches square, fired to Cone 05, Mary George Kronstadt, Washington, D.C.

Another Season, Another Palette

by Mary George Kronstadt

When I was preparing to enter Indiana University in the 1950s, my parents had only two suggestions for my future career direction: nursing or teaching. Instead, I majored in art history. Now, having happily made pottery my career for the past 25 years, I know that I chose the right path.

I touched clay for the first time in 1973 and was immediately hooked. After taking classes at Glen Echo Pottery in Maryland for several years, I branched out on my own, working in a tiny basement room equipped with an electric kiln and a kick wheel. As I was living in Washington, D.C., there was no chance I would be firing my work with gas or wood. At first, I worked with stoneware and porcelain at Cone 10, then ventured into low-fired majolica. It took eight terribly disappointing firings before I fired one moderately successful majolica-decorated piece.

Most of Kronstadt's work is for the table; forms are either wheel thrown or pressed in plaster molds.

Shortly thereafter, I noticed the new shopping center that was opening only two blocks from my house was going to include a craft gallery. I immediately called the owner, Jackie Chalkley; she agreed to carry a few of my first majolica pieces, and later feature my work in several solo shows. That was my commercial start.

From the beginning, I've also held an annual holiday sale in early December at my studio. Over the years, it has grown and changed, becoming an important part of my annual income. To build my client base, I have kept meticulous records. Having a seconds sale at the same time has had an impact as well. Customers line up, sometimes even in the rain, just to get first pick.

My studio is open all year by appointment. When customers visit, I pay attention to what they like and don't like, and sometimes develop ideas from what people say they want but can't find. I also take orders for dinnerware and personalized wedding platters.

Another important career move was to take time to serve on two interesting boards of directors—that of Pyramid Atlantic, a center for print, paper and books in Riverdale, Maryland; and also that of the James Renwick Alliance, a support group for the Smithsonian Institution's Renwick Gallery. This board work has opened a window on my whole profession—locally and nationally—as I have made contacts and learned immense amounts about every aspect of my craft. I have also taken craft trips all over the country, further expanding my exposure to interesting artists, important collections and people who might turn into clients.

My work is primarily for the table, although I have also done a number of tile installations in private homes. The best sellers are 12- to 14-inch pasta bowls. Next in popularity are butter dishes designed specifically to contain one stick of butter. Wine coasters and compotes have also been popular.

Teapot, 7½ inches in height, wheel-thrown and assembled, with brushed oxides and stains on majolica base glaze, fired to Cone 05.

Majolica-decorated platter, approximately 17 inches in length, press-molded red earthenware, Cone 05.

Platter, 21 inches in length, majolica-glazed earthenware, fired to Cone 05.

Decorating is by far the most time-consuming part of my production. I use commercial stains mixed equally with frit in water. Most of the designs are one-of-a-kind abstracts, with different color schemes and shapes. For example, the patterns on my large oval platters are based on fall leaves, applied using stencil and tracing techniques, layering color, sgraffito, line drawing and blending of colors.

The intense colors seen on a bicycling trip in Morocco last year inspired my new candlestick design. Sometimes, I try out color combinations seen in clothing, magazines, films and nature. I reuse some that I think are especially successful; otherwise, each piece is unique. The variety of designs keeps the work fresh for me.

I have also had the good fortune to have wonderful studio helpers. Currently, I hire graduate students from George Washington University. They have far more technical knowledge than I do; they have insisted that we redo a number of my plaster press molds, which were not up to their standards. I have learned a lot from them, and miss them as they graduate and go on to start their own ceramics careers. Then it is time for me to begin again myself—another season, another palette. ▲

Triangular vases, 14 inches in height, earthenware with majolica glaze.

Wheel-thrown covered jars, to 10 inches in height, majolica-decorated earthenware, by Mary George Kronstadt, Washington, D.C.

The Building Blocks of Electric Kiln Pottery—Clay and Glaze Formulation

Inspiration, a fertile imagination, and technical skill are necessary ingredients in the creation of any ceramic art—but without the proper materials, those assets are useless. A potter can't just use whatever clay and glaze is found lying around the studio and expect good results. And any pottery student quickly learns one can't just walk into a pottery supply store and ask for "clay and glaze." The modern electric kiln potter faces a mind-boggling array of choices, beginning with the choice of firing temperature, then continuing on to the types of clay and glaze, and whether to use commercial materials or to mix his own, whether to look for published glaze recipes or to formulate unique new glazes, and so on.

Certainly, many of the materials appropriate for electric kiln pottery are very different from those designed for other firing methods. A clay which is a warm, toasty color in reduction may be an unappealing yellow in oxidation. A seductive red reduction glaze is likely to yield an uninteresting green in oxidation. On the other hand, a deep red, iron-rich clay which fires to a lovely brick in an electric kiln may bloat in a reduction firing, and the rainbow of beautiful commercial glazes and stains formulated for oxidation may change drastically in another type of firing. It is necessary to experiment not only with new materials and recipes, but also to be sure those recipes and materials are designed for oxidation firing.

The previous chapters illustrated numerous approaches to electric kiln pottery and ceramic art, and include many clay and glaze recipes. Those recipes are merely a sample of the endless possibilities open to the electric kiln potter, however. The clays and glazes that are ideal for one potter may not be suitable for another's vision. Therefore, this chapter offers a closer look at the basic materials of the potter's art. Red clay is traditionally used for majolica and other electric kiln work, because the color is rich and appealing while white clays or porcelain may seem to have a "dead" look. On the other hand, the majority of commercial claybodies are white or buff—thus, there are a number of recipes for red clays here, along with information about formulating your own body recipe.

Though clay is the foundation of any ceramic art, many potters will be content with one or two clays and prefer to concentrate their research on the multitude of glaze surfaces possible. Materials for finishing the ceramic surface include slips and engobes, glazes based largely on clay or slip, and many types of glaze. This chapter offers several approaches to achieving an interesting glaze surface in a neutral firing atmosphere without relying on decoration applied either under or over the glaze. One successful approach involves layering glazes and slips, or different types of glaze, so that in addition to the interaction between clay and glaze the pot's surface benefits from the interaction between the materials it is coated with. Other potters prefer to use a single glaze with inherent visual texture, such as a crystalline glaze, or even to duplicate reduction glazes in the electric kiln.

Ceramists who mix their own glazes must also be prepared to deal with fluctuations in the supply of glaze ingredients. One ingredient that has long been the foundation of many valued oxidation glaze recipes—including many recipes in the preceding chapters—is Gerstley borate. When the supply of that material failed at the dawn of the twenty-first century, it threw the community of electric kiln potters into a frenzy of experimentation to find and formulate reliable substitutes. Though Gerstley borate is again available at the time of this writing, the supply is still questionable. Therefore, this chapter finishes with sugges-

tions for getting along without this material. When trying new recipes, it is always wise to test a small batch first, and thoroughly test the new glaze in interaction with your clay and other glazes before mixing a large batch. Any potter trying recipes from this book that call for Gerstley borate would be well advised to test substitutions for this material instead, and avoid any recipe which can't be successfully formulated without it.

Low-Fire Redware Bodies

By Jonathan Kaplan

Working with redware bodies in the Cone 06-04 range (approximately 1830°–1940°F) can be frustrating, particularly when compounded with a lack of specific "kiln-tested" information. One recipe may work for one potter, but it may not work for another in a different studio situation. The importance of doing one's own research and testing cannot be minimized.

While it is possible to purchase prepared plastic bodies and mixed casting slips for this low fire cone range, it is certainly not practical for a sizeable production. It is also difficult not knowing the formula of a manufactured body. If difficulties arise, and they will, it is close to impossible alter a prepared body, or "clay in the box." However, with research, reading, and testing, coupled with a small test kiln and controller, predictable and satisfactory results can be achieved when formulating your own clay body.

I have made redware, or terra cotta pots by throwing, Ram© pressing, jiggering, and slip casting. I developed two clay bodies. One, a plastic formula suitable for conventional hand forming techniques, and the second, a deflocculated composition for slip casting. Both bodies work well, and their fired characteristics are quite similar.

Both clay bodies utilize basically the same materials, but in different proportions and amounts. Cedar Heights Redart and any domestic ball clay could be used for the plastic content. The coloring ability of Redart is so intense that one could use even a small percentage and achieve a warm terra cotta color. By itself, Redart does not have very appealing working characteristics. It is quite sticky, quite dense, and difficult to work alone without any additions of other clays. Its properties in either a casting or plastic body are enhanced with the addition of either a ball clay or a kaolin.

Redart and most ball clays have very fine particle sizes and by themselves are very plastic. To achieve decent plastic or casting recipes, it is beneficial to cut the Redart content even further with the addition of coarser particle sized kaolins such as Tile 6, EPK, Pioneer, or Ajax P for the plastic body, and Velvacast for the casting body. There are specially blended ball clays specifically for slip casting such as Old Hickory FC-340. These clays allow for a more even drying body and help eliminate warping.

When jiggering bowls and plates, Jonathan Kaplan begins with a clay disk made from an extruded pug; turning the clay 90° eliminates bottom S-cracking (which sometimes results from the spiraling action of a pug mill's screw or even spiral wedging).

After the clay disk is pressed into the mold, piercing with a needle tool allows air to escape from the foot area.

The clay is smoothed into the revolving mold prior to defining the inner profile with the jigger template.

Sprayed water lubricates the clay as the jigger arm is pressed down into the revolving mold.

Excess clay is trimmed from the rim with a wooden tool; final trimming is done when the bowl is leather hard.

In a casting body, Velvacast promotes a faster casting rate.

It is important to understand what is called LOI or loss on ignition. Loss on ignition represents the percentage of weight lost from the burning out of chemically and physically combined carbonaceous materials such as lignites. The standard temperature range used to determine the weight loss is 1000°–1100°C or 1832°–2012°F and is achieved with a testing device called a dilatometer.

Because of the presence of significant amounts of lignite in both Redart and any ball clay, it is common practice to take the bisque firing higher than the glaze firing. A higher bisque of sufficient length in a well ventilated kiln will burn off the offending materials that can compromise the glaze surface. If this practice is not followed, the glazed surface sitting above the clay body is volatilized or parted out, leaving pinholes of varying types. One can see a rather dramatic demonstration of just how much combustionable material is contained within these clays by separately blunging some Redart and a ball clay. After mixing, pass the resulting slip through a 60 mesh screen. The screen residue is the LOI. This is the "junk", the offending materials that need to burn out cleanly before the ware is glaze fired. While sieving casting slips is part of the process of making slip, plastic clay bodies can also be mixed as a slip, sieved, and then dewatered with a filter press, passing finally into the prepug and then de-airing pugmill.

The use of feldspar as a flux and silica as a glass former is not practical in this temperature range as their melting point is much higher. Common low fire body fluxes are nepheline syenite, talc, or a leadless frit. Wollastonite, a calcium silicate is also a good addition to the clay body. It promotes even drying, reduces shrinkage, increases thermal shock resistance, and also increases the green strength of the body. However, wollastonite is hydroscopic. It absorbs moisture from the air and will form hard lumps in the bag. It is a good idea to mix this material thoroughly with water and sieve it prior to adding it to the clay body. Frits can be added to the clay body to promote vitrification. They are expensive, yet produce a very dense, durable body. If fired too high, the body develops too much glass and tightens up. Its absorbency is reduced to the point where it can no longer absorb the water from the glaze. The resulting glaze surface is then under developed.

A "happy medium" is easily arrived at by understanding the high fluxing potential or Redart and using talc and wollastonite with it in amounts that produce the desired absorbency and shrinkage. Such a body can be easily bisque fired to cone 04 and then glazed at cone 06 with good glaze surface development.

In any body containing significant amounts of iron bearing clays such as Redart, the presence of soluble salts in these materials can produce a scumming or efflorescence on the fired surface. This whitish film on the surface of the unglazed clay can be eliminated by adding a small percentage of barium carbonate into the formula. Using between .5 to 1% barium carbonate or barium sulphate will eliminate this defect. It is wise to mix the material with water first, screen it, and then add it into the plastic clay body as it mixes. In a casting body, keep the barium content very low, as the presence of a carbonate or a sulphate will tend to decrease the gelling of the slip, so necessary for casting. The blunging and screening of the slip mixture can easily disperse the barium compounds.

The difference between the casting formula and the plastic formula is that the casting recipe contains 50 parts plastic materials and 50 parts non-plastics. The body used for conventional hand forming methods, the plastic clay body, has a ratio of 90 parts plastic materials to 10 parts non-plastic materials. This ratio can be altered or changed to suit individual preferences. At this temperature range, body compositions are quite forgiving. The most important factor to remember when compounding any clay body at this temperature, or for that matter, at any temperature, is to make sure that the absorbency is low enough to produce a vitrified clay body that can survive daily use if you involved in producing usable functional ware. Both clay bodies are bisque fired to cone 04 and then glaze fired at cone 06.

The plastic body is fairly direct, using a combination of Kentucky ball clay (OM-4), EPK, Tile 6, and Redart. Talc (Nytal 100) and wollastonite (Vansil) are combined in equal amounts for added flux. The clay fires a bright orange, has approximately 8% shrinkage and is suitable for oven use.

High iron bearing clays such as Redart are problematic to deflocculate. Rather that using a sodium silicate and soda ash mixture that proved highly unsatisfactory, Darvan 811 or Spinks 211, both commercial polyacrylates prove to be very successful.

By testing small batches of slips it was only a matter of time to arrive at a casting body with acceptable working properties. The specific gravity works best at 1.75–1.77. With EPK as the only kaolin, the viscosity remained quite high in the 45–60 second range using a Lehman viscosimeter. One of the goals in formulating a casting slip is to have a low viscosity with a high concentration of solids. By adding Velvacast kaolin to the slip formula, the viscosity dropped to a very workable 23–25 seconds. This way, the casting rate can be adjusted to suit. Further, substitution of FC 340 for the OM 4 ball clay also had a beneficial effect on the casting rate. Kaolins are the weak link in a casting system. Ball clays impart strength and are the "glue" that holds the clay body together.

A good starting point for testing a casting body is to mix up 10 pounds of dry materials which yields approximately 1 gallon of slip. 50 ml of Darvan 811 is also a good starting point for the deflocculant. Slips can be cast in molds because of the gelling nature of deflocculated clay as well as its ability not to settle out. However, deflocculation changes on a daily basis because of the organic materials that are present in clays, especially high iron clays such as Redart. Test for specific gravity and viscosity prior to casting each time.

Red casting slips exhibit different deflocculation curves then their white clay counterparts, even if the specific gravity and viscosity are similar. So it is unwise to use these numbers to make a white slip. Substitution of any clay in the mix can alter these numbers which will alter the casting characteristics of the mixture. Most clay suppliers have adequate data and can be of great help in assisting with your formula.

Here are some general guidelines for mixing large batches of red casting slip. Use warm or hot water and begin by adding the deflocculant, barium and wollastonite and let them disperse in the blunger for approximately 5-10 minutes. Some add the plastic ingredients first, others add the non-plastic ingredients first. I add the non-plastics first and let them fully blunge, and then add the clays and blunge the entire batch thoroughly for at least a half hour. I then take initial viscosity and specific gravity readings and adjust to fit. The slip can blunge overnight or sit without overnight mixing. In the morning, test for viscosity and specific gravity, and make the final adjustments as necessary. Any slip, just like their plastic body counterparts will benefit by aging for a period of time before using.

Here are some formulas to test in your own studio:

Limit Recipes for Cone 04 Redware Bodies

	Light Firing	Medium Firing	Dark Firing
Nonplastic Kaolin	10-15%	0-15%	0-5%
Plastic Kaolins	20-40%	0-25%	0-15%
Ball Clays	5-25%	15-40%	10-30%
Stoneware Clays		15-40%	10-35%
Fireclays		15-30%	0-20%
Earthenware/ Red Clay		0-25%	40-80%
Lead-Free Frit	10-25%	0-20%	10-20%
Talc	10-20%	0-15%	0-10%
Wollastonite	0-10%	0-10%	0-10%
Nepheline Syenite	0-10%	0-10%	0-10%
Flint	0-15%	0-10%	0-10%
Grog		0-10%	0-10%

Redware Casting Body JZ1
(Cone 04)

Talc	10 lbs.
Wollastonite	10
Frit 3124 (Ferro)	5
Ball Clay	10
Cedar Heights Redart	65
	100 lbs

Add: Barium Carbonate 280g
Darvan 811 .... 500g

Redware Casting Body JZ 2
(Cone 04)

Talc	15 lbs.
Wollastonite	10
Ball Clay	10
Cedar Heights Redart	65
	100 lbs

Add: Barium Carbonate .. 280g
Darvan 811 .... 500g

Redware Casting Body CDG 2
(Cone 04)

Talc	25 lbs
Wollastonite	25
Redart	15
FC 340	15
EPK	5
Tile 6	5
Velvacast	10
	100 lbs

Add: Barium Carbonate .... 300 g
Darvan 811 .... 300 ml to start
Water .... 10 lbs to start

Redware Casting Body CDG 2A
(Cone 04)

Talc	22.5 lbs
Wollastonite	25.0
Frit 3124	2.5
Old Mine 4	5.0
FC 340	20.0
Redart	20.0
Velvacast	5.0
	100 lbs

Add: Barium Carbonate ... 300 grams
Darvan 811 .... 300 ml to start
Water .... 10 pounds to start

Redware Plastic Body CDG TC-1
(Cone 04)

Talc	5 lbs
Wollastonite	5
Old Mine 4	20
EPK	10
Tile 6	10
Redart	50
	100 lbs

Add: Barium Carbonate .... .5 lb

Redware Plastic Terra Cotta Body
(Andrea Gill Cone 04)

Wollastonite	10 lbs
Ball Clay	10
Redart	60
Fireclay	20
Fine Grog	10
	100 lbs

Add: Barium Carbonate .... 1/3 cup

Many other clays can be used in developing redware formulas. Ball clays such as Tennessee #5, Foundry Hill, and Spinks HC5 can be used and Goldart and screened fireclay can also be added.

Pyrophyllite is an exceptional material that can be added to these bodies to increase their strength. The addition of other non-plastics may lighten the fired characteristics, but with the exceptional coloring power of Redart it might be unnoticeable. Note that these formulas work in my facility under my conditions. Please test beforehand. They can also serve as a basic structure for experimentation with other ingredients in different quantities. ▲

Packing materials are stored in the loft above the office and machine shop in Kaplan's well-organized studio.

Right Redware vase with floral relief decoration, 8 inches in height, glazed on the inside, left unglazed on the outside, fired to Cone 06 in oxidation; a collaboration between Kaplan and JH Designs.

Below: Press-molded dinner plates with jiggered cup and bowl, decorated with stains brushed and sponge-stamped over Cone 06 glaze, by Jonathan Kaplan.

Red Clays
for Mid-Range Oxidation

by David G. Wright

Teapot, 9 inches in height, wheel-thrown and altered red clay, by David G. Wright, Moorestown, New Jersey.

My interest in red-burning clays was piqued during graduate school at the University of Colorado in Boulder. In fact, the word "Colorado" is Spanish for "red or reddish." The landscape of the Southwest has always held a tremendous fascination for me, and I knew it would affect my artwork, but I was unsure of how or when.

Several years later, I found myself living in New Jersey to be close to my family, but longing for the desert Southwest. This desire led to the development of clay bodies that provided the color, texture and feeling of that region.

I wanted a throwing body that was deep red when fired to Cone 6 in oxidation. To get a red at high temperature, I knew I needed a primary clay that was reddish, so I tried Cedar Heights Redart, but it only gave browns. I then found a clay called Neuman Red, which is mined in Sacramento, California, by H. C. Muddox. Neuman is bright orange in color and very refractory, having a PCE (pyrometric cone equivalent) of 19–20. Although the company considers it a ball clay, as 41.7% of its particle size is less than 200 mesh, the dry clay feels toothy like a fireclay.

Using a combination of Neuman with Redart, I quickly achieved the results I wanted. Neuman provided the color, while the Redart allowed for the Cone 6 firing temperature.

About this same time, my friend and former teacher Bill Daley was also in search of a red clay. He had previously (in 1973) developed a red stoneware body using PBX Valentine Fireclay, as a result of a commission he had received from the Ritz Theater in Philadelphia. Unfortunately, PBX is no longer being mined, so he had to reformulate his clay.

I began using Neuman Red in 1993 and introduced Bill to it soon after. We each conducted a lot of tests and shared our information, eventually developing the D/W body listed below. Bill ultimately adjusted this body again, using more Redart clay for greater vitrification and a darker color.

I went on to develop a palette of different reddish colored clay bodies, based on the Neuman and Redart combination. Along the way, I came up with not only red, but orange, tan, brown, chocolate and maroon bodies:

D/W Red Body
(Cone 6)

Ball Clay	10%
Cedar Heights Redart	30
Fireclay	20
Neuman Red	40
	100%
Add: Barium Carbonate	1 %
Bentonite	2 %
Fine Grog	5 %

A red-burning stoneware that is better for handbuilding than for throwing.

Wright "Red" Stone Body
(Cone 6)

Ball Clay	10%
Cedar Heights Redart	30
Fireclay	25
Neuman Red	35
	100%
Add: Barium Carbonate	1 %
Bentonite	2 %

This is my current body. It has a nice color and is very forgiving.

Daley's Body
(Cone 6)

Ball Clay	13.33%
Cedar Heights Redart	40.00
Fireclay	20.00
Neuman Red	26.67
	100.00%
Add: Barium Carbonate	1.00 %
Grog	10.00 %

This is Bill Daley's current body; it is less red than mine, but tight as a drum at Cone 6 oxidation.

Bach's Red Body
(Cone 6–8)

Ball Clay	10%
Cedar Heights Redart	25
Hawthorne Fireclay	25
Neuman Red	40
	100%
Add: Barium Carbonate	1 %
Bentonite	2 %
Molochite	5–12 %

A beautiful orange-red, but not as tight as the others, with perhaps 3–4% absorption at Cone 6.

Maroon Red Body
(Cone 6)

Ball Clay	10%
Cedar Heights Redart	40
Neuman Red	40
Flint	10
	100%
Add: Barium Carbonate	1 %

A deep, rich maroon red, this recipe is extremely short, but good for sculpture; very tight at Cone 6.

David's Brick Red Body
(Cone 6)

Nepheline Syenite	10%
Ball Clay	10
Cedar Heights Redart	20
Neuman Red	50
Flint	10
	100%
Add: Barium Carbonate	1 %
Bentonite	3 %
Fine Grog	5 %

With 13% shrinkage and 3% absorption at Cone 6, this recipe is a little more red than Wright "Red" Stone, but not as plastic.

Tight Red Body
(Cone 6)

Ball Clay	33.33%
Cedar Heights Redart	33.34
Neuman Red	33.33
	100.00%
Add: Barium Carbonate	1.00%

Yields a deep reddish color. My best throwing body. Grog is optional.

"Dove Descending (Holy Spirit)," 18 inches in height, handbuilt stoneware, fired to Cone 6 in oxidation, unglazed, set in cherry pulpit, by William P. Daley, Elkins Park, Pennsylvania.

Platter, 18 inches in diameter, wheel thrown from Bach's Red Body, fired to Cone 6, by Laurence Bach, Mount Airy, Pennsylvania.

Chocolate Body
(Cone 1–4)

Ball Clay	20%
Blackbird Clay	20
Cedar Heights Redart	50
Fireclay	10
	100%
Add: Barium Carbonate	1%
Bentonite	2%

A fairly good throwing body with moderate dry strength. Fires to a milk-chocolate color, with about 2% absorption, at Cone 1 oxidation.

Red Throwing Body
(Cone 1–4)

Ball Clay	20%
Cedar Heights Redart	70
Fireclay	10
	100%
Add: Barium Carbonate	1%
Bentonite	2%

A fairly plastic body for throwing or handbuilding. Yields a red-brown, with 2% absorption, at Cone 1 oxidation. An addition of 10% nepheline syenite will increase vitrification, but darkens the color considerably.

#18 Red Body
(Cone 1–4)

Ball Clay	10%
Cedar Heights Redart	80
Flint	10
	100%
Add: Barium Carbonate	1%
Bentonite	2%

A smooth, short body for throwing. Yields a beautiful deep red at Cone 4.

David's Toast Body
(Cone 1–4)

Ball Clay	33.33%
Blackbird Clay	16.67
Cedar Heights Redart	33.33
Fireclay	16.67
	100.00%
Add: Barium Carbonate	1.00%
Bentonite	2.00%

The preceding recipe yields a high-iron, very workable toasty brown clay for throwing and handbuilding.

Tan Throwing Body
(Cone 6)

Ball Clay	41.67%
Cedar Heights Redart	25.00
Fireclay	33.33
	100.00%
Add: Fine Grog	5.00%

A nice all-purpose clay.

Toasty Stoneware Body
(Cone 6)

Ball Clay	33.33%
Cedar Heights Redart	33.34
Fireclay	33.33
	100.00%

A nice all-purpose toasty warm stoneware for electric firing. Grog is optional. ▲

Versatile Cone 06–6 Clays and Engobes

by Gerald Rowan

"Painting with Clay Sections," with slabs of Brick Red Clay Body, fired to Cone 06, by Gerald Rowan, Allentown, Pennsylvania.

With my work equally divided between sculpture, painting and clay vessels, I thought it would be best to standardize my kiln firings to two cone levels: Cone 06 and Cone 6. This would allow me to bisque earthenware and stoneware as well as glaze fire earthenware all at the same time—an efficient use of both kiln space and energy.

This move toward standardization required clay bodies in a variety of colors and at both temperature ranges. Since I make clay vessels, fracture them and reassemble them dry for sections of my paintings, I also needed a series of slips that would allow me to join dry clay without much cracking. In order to join dry pieces of clay, I had to fit the joints well and slip generously with a low-shrinkage slip.

I found that the following clay bodies, slips, engobes and stains work well over a wide range of temperatures and are forgiving enough to be used for large vessels and sculpture:

Brick Red Clay Body
(Cone 06–6)

Ball Clay	13.80 lb
Cedar Heights Goldart	34.50
Cedar Heights Redart	34.50
Medium Grog	7.00
Fine Grog	7.00
Barnard Slip	2.51
Red Iron Oxide	0.69
	100.00 lb

For large pieces, medium and coarse grog may be substituted for the medium

grog content. Fine potters' sand may be substituted for the fine grog.

Brick Red Joining Slip
(Cone 06–6)

Cedar Heights Goldart	39.88%
Cedar Heights Redart	39.88
Calcined Kaolin	15.96
Barnard Slip	2.90
Red Iron Oxide	0.80
Sodium Silicate	0.58
	100.00%

Mix with only enough water to produce a thick slip.

Mid-Range Sculpture Body
(Cone 6)

Ball Clay	45.5%
Fireclay	45.5
Fine, Medium or Coarse Grog	9.0
	100.0%

Grog content may be increased to about 20% for large work.

Mid-Range Joining Slip
(Cone 6)

Calcined Kaolin	49.73%
Fireclay	49.73
Sodium Silicate	0.54
	100.00%

Mix with only enough water to produce a thick slip.

Low-Fire Sculpture Body
(Cone 06)

Ball Clay	45.5%
Talc	45.5
Grog or Sand	9.0
	100.0%

Use Number 1 silica sand or porcelain grog to keep this body white.

Low-Fire Joining Slip
(Cone 06)

Calcined Kaolin	49.73%
Ceramic Talc	49.73
Sodium Silicate	0.54
	100.00%

Mix with only enough water to produce a thick slip.

I also found the following stain and engobes useful for sculpture and decorative vessels; these are applied to bisqueware and fired in a well-ventilated kiln:

Black Stain
(Cone 06–6)

Frit 3110 (Ferro)	15.0%
Black Copper Oxide	4.0
Cobalt Oxide	1.0
Manganese Dioxide	40.0
Red Iron Oxide	40.0
	100.0%
Add: Bentonite	3.0%
CMC	1.0%

Glossy Engobe 200
(Cone 06–04)

Whiting	8.9%
Custer Feldspar	11.1
Frit 25 (Pemco)	50.0
Kaolin	9.0
Flint	21.0
	100.0%
Add: Bentonite	3.0%
CMC	1.0%

Matt Engobe 201
(Cone 06–04)

Ball Clay	16.0%
Whiting	9.0
Frit 25 (Pemco)	15.0
Kona F-4 Feldspar	50.2
Kaolin	9.8
	100.0%
Add: Bentonite	3.0%
CMC	1.0%

For each engobe recipe, dry mix all ingredients; add water and allow to stand overnight. If necessary, readjust water content before using.

Color variations of the engobes are possible with the addition of 10–15% commercial stain or the following oxide/stain combinations:

Enamel White:	
Tin Oxide	5.0%
Zircopax	7.5%
Majolica White:	
Tin Oxide	10.0%
Off White:	
Titanium Dioxide	10.0%
Medium Blue:	
Tin Oxide	4.0%
Cobalt Oxide	2.0%
Opaque Blue:	
Tin Oxide	5.0%
Cobalt Oxide	2.5%
Blue Gray:	
Tin Oxide	1.0%
Cobalt Oxide	0.5%
Iron Chromate	3.0%
Blue Green:	
Tin Oxide	4.0%
Black Copper Oxide	2.0%
Chrome Oxide	3.0%
Turquoise:	
Cobalt Carbonate	0.5%
Copper Carbonate	3.0%
Fluorspar	5.0%
Warm Gray:	
Cobalt Oxide	0.5%
Manganese Carbonate	0.4%
Pink Stain	1.5%
Dark Brown:	
Tin Oxide	2.0%
Manganese Dioxide	3.0%
Red Iron Oxide	5.0%

My preference is to work with vitreous engobes rather than glazes, primarily because engobes can be applied much like paint. Usually two coats applied with a soft brush yields good coverage. I also find I have a greater degree of control with a brush, as opposed to dipping or pouring; however, the above recipes could easily be dipped or poured to meet others' working habits or individual tastes. ▲

"Nar~ow-Necked Vessel," 18 inches in height, wheel-thrown porcelain, with brushed and trailed slip glazes.

Porcelain Slip Glaze

by Joseph Godwin

During a summer spent in picturesque Switzerland, I dreamed of painting impressionist landscapes, sunlit fields of golden rape-seed flowers and chocolate-brown wooden houses surrounded by bright red geranium flower boxes. But it rained every day, all day. Instead, I painted psychological portraits of a wet landscape and read C. G. Jung books on psychology. My paintings emerged as abstract, inner landscapes—nonobject and nonrepresentational.

Painting in Switzerland was a welcome change from many years of porcelain carving at my studio in Massachusetts. While in Switzerland, I also visited several potteries around the country; at one of these, the potters were developing stoneware slip glazes and I gave them the recipes for the porcelain slips I had developed for slip carving and inlay. By the time I left Switzerland, they had developed a series of opaque, stoneware slip glaze colors.

On returning to the States, I decided to continue working with the painting techniques I had studied that summer, but using porcelain slip glazes on porcelain vessels. I began by formulating a slip glaze with the same basic flux that the Swiss potters had used for their stoneware slip glazes—wollastonite. A natural calcium silicate, it is used to replace silica and whiting in clay bodies and glazes. My base test consisted of a combination of wollastonite and bone-

Porcelain vase, 18 inches in height, wheel thrown, brushed with porcelain slip glazes, fired to Cone 8.

When applying slip glaze decoration, Joseph Godwin balances the form on plastic-covered foam rubber.

dry Grolleg porcelain, equal parts by weight. The original 50:50 clay and wollastonite recipe, which had produced a matt stoneware slip glaze for the Swiss potters, fired to a semiopaque porcelain slip glaze when mixed with a commercially available, Cone 8–10, Grolleg porcelain body consisting of approximately 50% Grolleg kaolin, 25% G-200 feldspar and 25% flint, plus 2% VeegumT. I have found that 50 parts of the Grolleg porcelain body with 40 parts wollastonite fluxes as well as the 50:50 recipe.

Since so much of the slip glaze is a clay body, its original chemical and physical properties have a significant influence on the maturation temperature and application properties. A slip glaze formulated with a Cone 8–10 porcelain body will have a lower maturation temperature than one made with a Cone 9–11. A slip glaze with the proportions of 50 parts porcelain and 40 parts wollastonite will attain semiopacity or translucency when fired to the same cone (or slightly lower) as the porcelain body. The addition of 8%–10% Zircopax to the basic slip glaze will create an opaque white slip glaze; 4%–5% will yield a semiopaque slip glaze.

Lithium compounds in the form of petalite or spodumene (lithium feldspar) can be an important flux in porcelain slip glaze formulation. Lithium carbonate and lithium fluoride are also potential flux additives. They extend a slip glaze's firing range and in some cases help to control crazing in translucent slip glazes. Wollastonite itself has the property of reducing shrinkage in clay bodies and glazes, thereby preventing crazing problems in the opaque white slip glaze. When combined with Zircopax, it prevents crazing in the Opaque White Slip.

A lithium compound combines well with a frit of a low-fire feldspar, such as nepheline syenite, in a translucent slip glaze. Nepheline syenite contains a large percentage of soda and potassium in proportion to its alumina and silica content. This composition categorizes it as a low-fire soda spar. It fluxes well with a lithium compound, such as spodumene. Approximately 5% frit or nepheline syenite, combined with 5% petalite or spodumene in a slip glaze, can expand the vitrification range of a Cone 8–10 slip glaze to Cone 6. These fluxes also afford stronger color saturation of colorant oxides and stains.

The basic purpose of a porcelain slip glaze is to facilitate the glazing of both green- and bisqueware with user-friendly versatility. Relatively little had been accomplished in this vein until the arrival of modern deflocculants and drying agents, which keep the liquid slip glaze in suspension for application purposes and allow it to dry correctly on bisque, thus preventing crawling during the glaze firing.

I prefer to use Veegum T for slip glaze suspension. It is a processed, colloidal magnesium alumina silicate that is used as a plasticizer in commercial porcelains. It consists of 80% Veegum, an inorganic bentonite, and 20% CMC gum, an organic binder. The colloidal property of 2% Veegum T significantly increases plasticity in porcelain clay bodies. The presence of 1% Veegum T creates an excellent deflocculant for porcelain slip glaze; it causes a mild thixotropic reaction. The addition of 1% bentonite increases thixotropy.

Frequent stirring during glazing with a porcelain slip glaze that includes 1% Veegum T is unnecessary. It takes several hours for a porcelain slip glaze containing Veegum T to complete a colloidal, mild thixotropic reaction, during which time a thin film of water forms on the surface as the slip glaze gels into a suspension rather than settles to the bottom of the bucket as glazes tend to do. When shaken or stirred, the slip glaze returns instantly to its former liquid state. It is therefore important to mix dry slip glaze recipes with a measured quantity of water, not only to ensure the correct consistency for your particular application purpose, but also to guarantee the correct consistency for a thixotropic glaze suspension.

If the slip glaze is mixed too thin, excess water will cause an uneven suspension, rendering the mixture unsuitable for glazing. The excess water will hold only the finer slip glaze particles in suspension, while most of the slip glaze forms a stiff mass at the bottom of the bucket. Excess water might not be decantable, without removing some of the finer slip glaze particles, for several weeks in a large volume of slip glaze.

As a safeguard, I measure the correct volume of water for a fairly thick slip glaze solution, then carefully thin the slip glaze to the desired consistency during and after sieving. The ratio of 1 cup of water per 10 ounces of dry slip glaze mix creates a consistency of slip glaze thick enough to brush and thin enough to pass easily through a 100-mesh sieve. Because a slight amount of water can significantly alter the viscosity of a small volume of liquid slip glaze, I thin a cup or two of thick slip glaze with squirts of water from an ear syringe.

PHOTOS: MIKE NESMITH

"Little Round Vessel," 5Z\v inches in height, with Cerulean Blue beneath Delft Blue, Golden Yellow, Orange and Red Porcelain Slip Glaze.

A drying agent is useful to ensure even drying, which in turn prevents thick layers of slip glaze from lifting from the bisque surface during application. Stain colors containing metallic oxides tend to increase the surface tension of a slip glaze. This impedes adhesion, resulting in the drying slip glaze's tendency to crack and peel away from the bisque surface. Glycerine works very well as a drying agent, especially for successive layers of slip glaze colors involving variable thicknesses applied onto moist bisque. The addition of 1 tablespoon of glycerine (6.25% fluid volume) per 1 cup of slip glaze is adequate.

Each time I dig into a pile of bone-dry porcelain shavings beneath the trimming wheel, I feel as if I am rediscovering the concept of glaze. To mix a porcelain slip glaze base, I add an equal weight of wollastonite to the bone-dry porcelain, or four parts wollastonite (by weight) to five parts bone-dry porcelain, depending upon which slip glaze base is desired. First, I dry mix outdoors, then pour the mixture into a container of steaming hot water premeasured by volume. The hot water quickly slakes the bone-dry scraps of porcelain, and the slip glaze is ready to pass through a 100-mesh screen within minutes. That's all there is to mixing a container of porcelain slip glaze from scraps, as the correct proportion of Veegum T is already included in the porcelain clay body. I need only add small proportions of opacifiers or fluxes for translucency, then the colorant(s) of choice, and glycerine for application.

Either base slip glaze combines well with commercial stains, but it is important to test each color, because some commercial stain colors are elusive at high temperatures. To mix a color test, I add the percentages of oxides and stains for a particular color to 10 ounces of thick liquid slip glaze base (prepared by mixing 10 ounces of dry slip glaze with 1 cup water) and resieve.

The above ratio of water to dry slip glaze creates a thickened slip glaze ideal for brushwork on greenware and bisque. An additional 20% water (by volume) thins the glaze enough for dipping or pouring on bisque. Adding glycerine as a drying agent is unnecessary for dipping and pouring. I mix a large liquid volume

Porcelain bottle, 13 inches in height, with layered porcelain slip glazes, by Joseph Godwin, Pepperell, Massachusetts.

of each slip glaze base, but add the glycerine only to small containers of colored slip glaze, after I mix in the colorants and resieve.

Three porcelain slip glaze bases that I presently use yield opaque, semiopaque and translucent results:

Opaque Slip Glaze Base
(Cone 8–10)

G-200 Feldspar	13.9%
Wollastonite	44.4
Grolleg Kaolin	27.8
Flint	13.9
	100.0 %
Add: Zircopax	8.0 %
Bentonite	1.0 %
Veegum T	1.0 %

The opaque recipe fires to a white gloss at Cone 8. The addition of 8% Zircopax is optional when formulating opaque colors with some stains, such as reds and yellows, because they contain opacifiers; 4% Zircopax is often sufficient for opacity.

Semiopaque Slip Glaze Base
(Cone 8–10)

G-200 Feldspar	11.9 %
Petalite (or Spodumene)	4.8
Wollastonite	47.6
Grolleg Kaolin	23.8
Flint	11.9
	100.0%
Add: Bentonite	1.0%
Veegum T	1.0%

The semiopaque base fires to a white semimatt on porcelain at Cone 8, but will become translucent at Cone 10.

Translucent Slip Glaze Base
(Cone 8–10)

Wollastonite	42.1%
Frit 3269 (Ferro) or Nepheline Syenite	5.2
G-200 Feldspar	13.2
Grolleg Kaolin	26.3
Flint	13.2
	100.0%
Add: Bentonite	1.0%
Veegum T	1.0%

The translucent slip glaze recipe utilizes frit and petalite additions to create translucency at Cone 8. The presence of lithium helps to prevent pinholes from forming in slip glazes containing frit and stains.

Some frits and stains reach their maximum glaze-use temperature below Cone 10. A translucent slip glaze containing these may begin to boil and form pinholes unless it contains a percentage of lithium to extend the recipe's maturation temperature range.

When mixed with trimming scraps from a Cone 8–10 Grolleg porcelain body, the recipes are as follows:

Opaque Porcelain Slip Glaze
(Cone 8–10)

Wollastonite	4 oz.
Zircopax	1
Bone-Dry Porcelain Body	5
	10 oz.

Semiopaque Porcelain Slip Glaze
(Cone 8–10)

Petalite (or Spodumene)	0.5 oz.
Wollastonite	5.0
Bone-Dry Porcelain Body	5.0
	10.5 oz.

Translucent Porcelain Slip Glaze
(Cone 8–10)

Petalite (or Spodumene)	0.5 oz.
Wollastonite	4.0
Frit 3269 (Ferro) or Nepheline Syenite	0.5
Bone-Dry Porcelain Body	5.0
	10.0 oz.

Mix each of the above scrap-clay recipes with 1 cup hot water, then add 1 fluid tablespoon glycerine for brushwork on bisqueware.

Porcelain slip glazes formulated with

oxide and stain additions, and fired in oxidation can achieve maximum color saturation. Muted color tones and textures can also be achieved by layering Volatile Black Slip Glaze beneath opaque white and colored slip glazes:

Volatile Black Slip Glaze
(Cone 8)

Hardwood Ash	4.17%
Wollastonite	18.75
Temmoku Glaze	52.08
Bone-Dry Porcelain Clay	25.00
	100.00%
Add: Black Stain	4.17%

Temmoku Glaze
(Cone 8)

Gerstley Borate	10.87%
Whiting	6.52
G-200 Feldspar	76.09
Kaolin	4.35
Flint	2.17
	100.00%
Add: Red Iron Oxide	8.70%

The volatile black melts at a slightly lower temperature and bleeds through the outer layers to create a mottled surface similar to reduction-fired stoneware in which iron particles in the stoneware clay speckle the surface. Varying proportions of oxides and stains have a significant effect on glaze melt and surface texture according to the flux capability of each colorant. Black colorants tend to have a strong fluxing quality.

Black glossy and black matt slip glazes can be differentiated with a slight alteration in the ratio of whiting to flint in the form of wollastonite. The following example substitutes half of the wollastonite with whiting in the glossy black recipe to create a black matt:

Glossy Black Porcelain Slip Glaze
(Cone 8–10)

Wollastonite	42.86%
Bone-Dry Porcelain Body	57.14
	100.00%
Add: Zircopax	7.14%
Black Stain	7.14%

Matt Black Porcelain Slip Glaze
(Cone 8–10)

Whiting	21.43%
Wollastonite	21.43
Bone-Dry Porcelain Clay	57.14
	100.00%
Add: Zircopax	7.14%
Black Stain	7.14%

A satin matt slip glaze can be calculated by altering the ratio of whiting and flint found in the base recipe. To alter a translucent slip glaze containing a frit into a satin matt, a ratio of 30% whiting to 10% wollastonite is introduced. The following satin matt slip glaze fires to a satin texture that works especially well with red colorants:

Satin Matt Porcelain Slip Glaze
(Cone 8)

Petalite	5.0%
Whiting	30.0
Wollastonite	10.0
Frit 3269 (Ferro)	5.0
Dry Porcelain Clay	50.0
	100.0%

A full palette of porcelain slip glaze colors is the most important part of my porcelain slip glazing process. Complex color combinations can be created by layering translucent colors over opaque and semiopaque colors so that they flow and pool. For the following color variations, add oxides and/or stains as specified to 10 (or 11) ounces of liquid porcelain slip glaze.

Ruby Red:

Mason Stain 6001	6.00 grams
Mason Stain 6003	6.00 grams
Mason Stain 6006	6.00 grams
Mason Stain 6031	6.00 grams

Peach Bloom:

Ferro Pink Stain	18.00 grams

Cinnabar Red:

Ferro Pink Stain	12.00 grams
Reimbold & Strick Stain K2323*	12.00 grams

Orange:

Ferro Pink Stain	6.00 grams
Reimbold & Strick Stain K2323	6.00 grams

Golden Yellow:

Reimbold & Strick Stain K2323	25.00 grams

Yellow:

Reimbold & Strick Stain K2323	12.00 grams

Aqua Green:

Mason Stain 6201	3.00 grams
Mason Stain 6364	6.00 grams

Deep Green:

Mason Stain 6202	3.00 grams
Mason Stain 6263	6.00 grams

Cerulean Blue:

Cobalt Carbonate	0.25 grams
Copper Carbonate	1.20 grams
Mason Stain 6364	18.00 grams

Sky Blue:

Mason Stain 6363	3.00 grams
Mason Stain 6364	6.00 grams

Turquoise Blue:

Mason Stain 6390	25.00 grams

Cobalt Blue:

Cobalt Carbonate	6.00 grams
Red Iron Oxide	3.00 grams

Delft Blue:

Copper Carbonate	3.00 grams
Red Iron Oxide	1.50 grams

Violet:

Mason Stain 6319	6.00 grams
Mason Stain 6385	6.00 grams
Reimbold & Strick Stain K2323	6.00 grams

Recent tests have resulted in a Super Opaque Porcelain Slip Glaze that gives excellent results over Delft Blue as well as Glossy Black:

Super Opaque Porcelain Slip Glaze
(Cone 9)

Wollastonite	33.33 %
Zircopax	11.11
Bone-Dry Porcelain Body	55.56
	100.00 %

For color variations, try adding up to 10% stain.

The original opaque recipe does not hold a strong white when applied over blue and black slip glazes. However, I continue to use it to layer between colored slip glazes and to glaze the insides of vessels.

Porcelain slip glaze has become a process for me to transform clay and glaze into color expression. Since I have chosen to immerse myself in this glazing technique, color composition has become the major theme of my work. Wheel-thrown vessels have become objects to radiate color, and the forms have begun to emerge from the wheel more free flowing. With porcelain slip glaze, ongoing work has gained momentum, developing a free-flowing connection between the greenware and bisque stages. ▲

---

*Reimbold & Strick stain is available from Fusion Ceramics, Post Office Box 127, 160 Scio Road, Southeast, Carrolton, Ohio 44615; (216) 627-2191.

Layered Cone 6 and Cone 06 Glazes

by Lana Wilson

PHOTOS: JOHN DIXON

"Ritual Teapot," 9 inches in height, with Lana's Red Moss Glaze, fired to Cone 06 in oxidation.

Continued interest in painterly textured surfaces and strong color has greatly influenced the surfaces of my ceramic work, but I haven't always been able to find the glazes that closely match my interests. Consequently, I recently went into partial seclusion and did about 800 test firings. I ended up working with five methods to develop certain glaze effects, and thought the specifics might be of interest and help to others. These methods are: 1) using simple ratios of glaze ingredients, such as 80:20 or 60:40; 2) adding fluxes to Cone 6 glazes to produce Cone 06 glazes; 3) testing fluxes with colorants; 4) altering existing glaze recipes; 5) adding ingredients by increments.

Layering also is an important technique for developing textured surfaces in electric firing. If a glaze is almost interesting, a different glaze under it can give it much richer surface interest. Layered applications of glazes can be fired at one temperature; or the first one can be fired at a higher temperature, then a second glaze applied and fired at a lower temperature.

As basic procedural information: I applied all of the glazes on white stoneware. (Since many of the ingredients—lithium carbonate, manganese, chrome, barium carbonate, etc.—are toxic, I always wear a charcoal-filter face mask and thin rubber gloves, and use basic safety precautions when mixing and applying glazes.) The Cone 06 glazes have also been successful on top of Cone 06 white slip on Cone 06 white clay.

The clay, water and glaze materials from different parts of the country sometimes make a very noticeable difference in results, as does a thin or thick application. By thick I mean ¼ inch thick. For thin, especially for the Purple Aqua, slightly thicker than milk is an appropriate consistency. Most of these glazes give one color when thin and a different color when thick.

Color and texture are also affected by how glazes are layered. If the first glaze is applied thickly, it will have more effect on the glaze layered over top. Usually I fire the first glaze application to Cone 6, then apply the second glaze and fire to Cone 06.

Although I wish I had dramatic surface-effect glaze recipes that were reliable every single time I fire, almost all of the glazes I like the best are temperamental. If you are interested in testing these glazes, try about five different application thicknesses. You may also need to adjust the recipes slightly for your ingredients and water, or fire at a cone above or below the one I suggest.

Using Simple Ratios

Lana's Red Moss Glaze illustrates the use of ratios to develop glazes. I knew lithium and manganese produced interesting brick reds, and when I was at Robin Hopper's glaze workshop in Canada, Brian Gartside suggested that I try simple 60:40, 70:30 or 80:20 ratios of lithium and various stable ingredients like kaolin and flint. This glaze was the result of that simple ratio approach. It is fired to Cone 06, but is usually layered over a lichen glaze fired to Cone 6, so I will give two lichen recipes first. These glazes can be dipped or brushed, but they are more successful if they are sprayed.

Lana's Lichen Glaze
(Cone 6)

Magnesium Carbonate	22%
Nepheline Syenite	60
Ball Clay	18
	100%

A white glaze. To cut shine add 5% increments of ball clay, kaolin or alumina. Apply thick and thin to test different results.

Mark Bell's Lichen Glaze
(Cone 6–8)

Magnesium Carbonate	31%
Talc	8
Zinc Oxide	6
Frit 25 (Pemco)	6
Soda Feldspar	30
Kaolin	19
	100 %

To achieve a cracked-earth effect, some cracks should show before this or the previous lichen glaze is even fired.

Lana's Red Moss Glaze
(Cone 06)

Lithium Carbonate	60%
Talc	40
	100%
Add: Bentonite	2%
Frit 25 (Pemco)	2%
Manganese Dioxide	6%

This is a simple ratio glaze. The manganese addition can be 3% for a light orange, 6% for a strong rich red and 10% for a darker red.

This is one of those temperamental glazes. It is very important to apply it thickly enough (up to ¼ inch) to yield red. A thinner application yields black. Spraying seems to give the best result, but I sometimes brush more on top to thicken some areas. For good color, I often do two applications and two firings of the Red Moss, but it seems better to do a thick enough application the first time than to have to refire.

To produce a brown variation, add 4% crocus martis or 2% Mason Stain 6600 instead of manganese. Another interesting brown can be made by using lithium carbonate 70 and talc 30. The bentonite and Frit 25 stay at 2%.

Another version of a glaze developed from a simple ratio is the following 80:20 chartreuse recipe. The Gerstley borate and flint together function as the 20% in the ratio. I started with all flint, but it was too dry so I gave 5% over to Gerstley borate to flux the mixture. I usually spray this thick for yellow and chartreuse, and thin for a purple gray. First I spray one of the lichen glazes listed above and fire it to Cone 6. Then I spray on Chartreuse Moss and refire to Cone 06. Occasionally the color changes over time after it has been fired.

"Southern California Ritual Teapot," 13 inches high, with Lana's Purple Aqua Glaze.

Lana's Chartreuse Moss Glaze
(Cone 06)

Gerstley Borate	5%
Lithium Carbonate	80
Flint	15
	100%
Add: Bentonite	2%
Chrome Oxide	3%
Tin Oxide	7%

This glaze yields a mild purple where thin, but often needs two or even three applications to get a strong lime color. Results are most interesting over Mark Bell's Lichen Glaze.

One variation I make requires no bentonite and uses kaolin instead of flint. This changes the color in a subtle way, and I sometimes layer both recipes on one pot.

Altering Cone 6 Glazes

My second method for developing glazes is the unconventional approach of firing intriguing Cone 6 glazes to Cone 06, adding enough flux to get them to work. The two that follow also happen to be based on simple ratios of glaze ingredients. Ceramics books often mention using nepheline syenite and barium carbonate with copper at Cone 6 to get striking colors. I tried this, but wanted a glaze to fire at Cone 06 and ended up altering the formula by adding a frit (in this case Ferro Frit 3289).

Lana's Purple Aqua Glaze
(Cone 06)

Barium Carbonate	45 grams
Nepheline Syenite	45
	90 grams
Add: Frit 626 (Pemco) or 3289 (Ferro)	4 grams
Bentonite	2 grams
Copper Carbonate	4–8 grams

This recipe should not be used on a food container, even on the outside. Spray thin for purple and thick for aqua. For more aqua, use smaller amounts of copper carbonate. The purple color can be varied by adding 4% to 8% copper Borax Engobe and spray Lana's Purple Aqua over it, then fire to Cone 06–05.

This glaze looks good sprayed or brushed over Dry Borax Engobe. Results are very dependent on the firing temperature and will change with only 50° difference. If it is too strong when it comes out of the kiln, try overspraying with Dry Borax Engobe in some places and refire, or apply a thin coat of Dry Borax Engobe and spray Lana's Purple Aqua over it, then fire to Cone 06–05.

Dry Borax Engobe
(Cone 06–6)

Borax	15%
Kaolin	55
Flint	30
	100%

For a speckled surface, add 3% granular rutile and brush on a thin coat. Sometimes if it is too thick, it flakes off. If this happens, add 2%–5% frit.

I often fire glazes at both Cone 06 and Cone 6 to see what will happen. Even commercial Cone 06 to 04 glazes can be overfired with good results to Cone 6 and some to Cone 5, though some run excessively. Blues, greens and blacks remain at these higher temperatures, but yellows, oranges, etc., usually do not. The following glaze was developed in relationship to the Aqua Purple:

Lana's Bronze Aqua Glaze
(Cone 6)

Strontium Carbonate	50.0%
Nepheline Syenite	50.0
	100.0%
Add: Gerstley Borate	2.5%
Bentonite	3.0%
Copper Carbonate	4.0%

A thin application produces aqua; whereas thick yields bronze. It runs wherever it's thick, so use a thin application near the bottom. A better bronze is produced if it is applied over a thin application of Bell's Lichen Glaze.

Flux Color Tests

The third method of developing glazes is a simple series of tests to determine which flux yields the most interesting results with a chosen colorant. To figure out which flux I should use with vanadium to make a rich rust, I combined several appropriate Cone 06 fluxes (Gerstley borate, frits, borax) at 20 grams to 2 grams of vanadium pentoxide. Borax produced by far the most interesting color. Then the borax and vanadium were developed into a workable volume-measured glaze with my most commonly used stabilizers: kaolin and flint.

Lana's Rust Glaze
(Cone 06–6)

Borax	1 teaspoon
Kaolin	½ teaspoon
Flint	½ teaspoon
Vanadium Pentoxide	¾ teaspoon
	2¾ teaspoons

Lana's Rust can be layered on itself, fired first at Cone 6, then more glaze brushed on and fired at Cone 06. It is not good over Bell's Lichen Glaze. There are also changes in the glaze colors if it is mixed and then applied days later because the soluble borax breaks down. For more rust, go heavier on the vanadium. This, of course, is quick to mix because it uses volume not weight.

This is definitely not my most popular glaze. People sometimes make a point of saying how much they dislike it.

One of my students, Paul Holden, has done this basic color response test with cobalt at Cone 8 in gas and electric kilns. First he tested several of the common feldspathic family of fluxes (Cornwall stone, soda feldspar, potash feldspar, nepheline syenite, spodumene, lepidolite) with 0.5% cobalt carbonate. This basic testing showed him how mottled and purple cobalt can be in lepidolite. Then he added secondary fluxes (whiting, dolomite, magnesium carbonate, zinc oxide, titanium dioxide, barium carbonate, strontium carbonate, etc.) to the lepidolite. In a test with titanium (5 parts lepidolite and 2 parts titanium), 0.5% cobalt carbonate turned green. The titanium pushed the cobalt to green. Next, he needed to stabilize or fill out the glaze formula with other ingredients like flint and kaolin.

This is a very visual way of learning the influence of basic glaze ingredients on colorants. It becomes clear that the effects of specific feldspathic fluxes and secondary fluxes can push a color like cobalt from green to blue or to purple.

Altering Existing Recipes

The fourth method of developing a glaze is simply to alter an existing glaze. I actually made the following recipe more unstable than the original, but then all the glazes I tend to be interested in are temperamental and tricky to use.

I like what lithium does to blue and have tested almost every lithium glaze I have come across. I found the following recipe in several places and wanted to make it drier. In the process of making it drier or less shiny, I also made it somewhat unstable.

Lana's Almost Stable Blue Glaze
(Cone 06)

Lithium Carbonate	27 grams
Bentonite	3
Kaolin	21
Flint	48
	99 grams

For blue green, add 4–5 grams copper carbonate; for cobalt blue, add 2 grams cobalt carbonate.

Apply a thin layer of Bell's Lichen, fire to Cone 6, then spray on Almost Stable Blue and fire to Cone 06. One firing seems more stable than two, but the color can be better with two applications and two firings. Try the cobalt version on top of the copper carbonate version and by itself.

Adding Incremental Amounts

My favorite and the easiest way to alter glazes is incremental testing. I would

"Artifact Teapot," 9 inches in height, with Lana's Bronze Aqua Glaze.

try this approach first if you only have the time to pursue one method of glaze development. To a black engobe that was too matt, I added 10% increments of nepheline syenite to produce a slight sheen in the surface. After each 10% was added, I mixed it up, brushed it on the test tile, then added the next 10% to the same cup.

Work with a minimum of about 50 grams when doing two to five incremental additions in one cup. It is slightly inaccurate for the last tests because some glaze has been used up in the previous applications, but you'll get a general idea and more accurate testing can be done later. I usually add at least 20% above what I think I will need because I see a greater range of possibilities and thus learn more.

Incremental additions can be done with the feldspar, starting lower than the suggested amount and adding till you go above the original formula amount. It is also intriguing to add secondary fluxes, such as zinc oxide, barium carbonate, bone ash, lithium carbonate, etc., in increments.

John Conrad's White Engobe
(Cone 6)

Borax or Soda Ash	5%
Whiting	5
Nepheline Syenite	10
Potash Feldspar	15
Kaolin	20
Kentucky Ball Clay	20
Flint	25
	100%

"Ten Legs Teapot," 10 inches in height, with Lana's Almost Stable Blue Glaze.

"Ritual Teapot with Hobo Symbol," 5 inches in height, with Altered Conrad Engobe.

Altered Conrad Engobe
(Cone 6)

Borax or Soda Ash	4%
Whiting	4
Nepheline Syenite	26
Potash Feldspar	12
Kaolin	17
Kentucky Ball Clay	17
Flint	20
	100%

I brush the Altered Conrad Engobe on a textured white clay, then sponge it off to create a black-and-white effect. If applied too thickly, it will pinhole. For either version, make black by adding 10% Mason Stain 6600. To make the engobe very dry for other effects, drop the nepheline syenite completely from the original formula.

Final Thoughts

A test kiln with a timer and kiln sitter is the one tool I should have bought earlier on, because I learn faster when I can test more often and follow an idea through. Most glazes seem to take about ten alterations before I get something I want to use.

Over the course of the testing, if I decide to produce an iron-related glaze using rutile, crocus martis, Cedar Heights Redart or Barnard clay, I use titanium dioxide, tin oxide, bone ash, borax, Pemco Frit 54 or spodumene to enhance color development. If I want to use either copper or cobalt as colorants, then I work with barium carbonate, strontium carbonate, lithium carbonate, borax, bone ash, Pemco Frit 626 or Ferro Frit 3289 and magnesium carbonate to encourage blues, purples and greens.

Layering offers a lot of potential for surface and color development. Sometimes I even cover up a failed effort partially or totally with a dry engobe and try all over again. And therein lies the secret of ceramic glaze development: try again! ▲

The Fugitive Blue Chun

by Emman Okunna

Perhaps as many varied descriptions have been given of Chun (Jun) glazes as there are potters. Frank Hammer in *The Potter's Dictionary of Materials and Techniques* speaks of Chun as "a feldspathic pale blue opalescent glaze on a buff body." However, in *Oriental Glazes*, Nigel Wood disagrees, noting that they are not correctly termed feldspathic: "Only about 30% of feldspar would have been needed to provide the 5% to 6% of potash and soda found in Northern (China) Chuns. Even less would suffice, as most Chuns show a high P_2O_5 content typical of glazes containing ash—and most ashes contain some potash. This means that the Chun glazes are not feldspathic in the strictest sense, since 50% feldspar is usually held to be the lower limit for feldspathic glazes. Wood also suggests that the blue opalescence of Chuns is favored by phosphorous, lead oxide and boric oxide from colemanite.

In *Pioneer Pottery*, Michael Cardew recognizes that "the opalescent colors of Chun glazes...and their causes remain a matter of conjecture." He believed they are optical colors produced by what he described as a "suspension of liquid in liquid (or rather glass in glass)." According to Cardew, wood ash, talc and some sort of frit are thought by some potters to favor Chun colors.

Daniel Rhodes (*Clay and Glazes for the Potter*) suggested that a glaze containing 4% to 8% bone ash and about 1% or less iron oxide will yield a bluish opalescent color if fired in a heavily reduced atmosphere.

In all these descriptions the common denominator has been the blue opalescence, a quality that characterizes and distinguishes the Chuns from among all Oriental ash glazes.

The following results of tests conducted at Nnamdi Azikiwe University in Awka, Nigeria, indicate several possible ways by which Chun glazes could be developed.

Empirical Approach

As Cardew aptly reasoned, "Since the theory of these colors is still uncertain, any rules on how to produce them must be purely empirical, based on observation and on evidence."

Our first consideration was that the materials from which ashes would be made should be easily available and in fairly large quantities to provide a consistent source. Ashes were therefore made from rice husks and corncobs, which were collected from nearby Achalla and Ugwuoba. Large quantities of kitchen ash were collected from a bakery here in Awka as well.

All ashes were washed four times and passed through 60-, 80- and 100-mesh screens. They were then dried, labeled and stored for use.

Biaxial and triaxial blends were mixed and fired to 1280°C (2336°F) in both oxidizing and reducing atmospheres.

Making Blue Chuns

There would be little use in giving a complete description of all tests carried out, as it will perhaps be sufficient to indicate the limits or amounts of materials used. Our first recipes were based on Katharine Pleydell-Bouverie's recipe of 40% ash, 40% feldspar and 20% clay. This recipe was modified slightly to bring in some minor ingredients—clay to help keep the glaze materials in suspension; talc ($3MgO{\cdot}4SiO_2{\cdot}H_2O$) to introduce the magnesium oxide (MgO) necessary for producing blues, and Enugu sand (levigated from Enugu clay) to introduce more silica (SiO_2) into the glaze batch.

Our first significant result was obtained by introducing both hard and soft ash into the following recipe:

Chun Glaze Test W
(Cone 10)

Rice Husk Ash	40%
Kitchen Ash	20
Talc	10
Potash Feldspar	30
	100%
Add: Sieved Enugu Sand	5%

The glaze melted, producing a rather strange yellow Chun, with a very weak blue—an indication of a high lime glaze. However, the result represented a remarkable improvement over earlier tests, because for the first time blue color was evident. We suspected at this point that two factors may have been responsible for nonachievement of blue color all along: the amount of feldspar used may have been quite high (40%); and the Enugu sand may still have had clay in it and so invariably increased the amount of clay in the recipe. But there was still no way of telling the exact amount of these materials.

A similar recipe (marked Z) in which kitchen ash was substituted with corncob ash was made up and blended against recipe W using the "eleven blend" biaxial method. The tests were applied over an iron glaze (temmoku)—a practice Cardew suggested would have the effect of bringing the Chun to life.

The results largely indicated that the corncob ash is more siliceous than the kitchen ash, as most of the tests were underfired where they were applied thickly. Only the 50:50 mixture of the two recipes showed a comparatively stronger blue. These results and a further reappraisal of the materials led to the decision to use both corncob ash and kitchen ash in the same glaze batch (as had been done with rice husk ash and kitchen ash), which yielded the following recipe:

Chun Glaze Test AF
(Cone 10)

Corncob Ash	40%
Kitchen Ash	20
Talc	10
Potash Feldspar	30
	100%
Add: Sieved Enugu Sand	5%

This recipe was also blended against recipe W, and the tests again applied over an iron glaze. The results were most exciting, and represented a breakthrough. All 11 tests produced dense Chun glazes, remarkably including the one that was the same as test recipe W, which had earlier produced only a weakish blue. However, the glazes were fluid. When the glaze was applied over an iron slip (deliberately made rich in iron and silica but low in alumina), instead

Wheel-thrown stoneware vase, 4 inches in height, with Chun glaze over iron-rich slip, fired to Cone 10 in oxidation.

of the temmoku glaze, the running was cured. It had also become bluer.

It is perhaps noteworthy that duplicates of these tests were also fired in reduction, and these results were remarkable, too. The tests that had produced blue colors in the oxidation firing now turned out fluid white-blue gloss Chuns with yellow specks. We suspect these specks to be lime crystals, as they appeared more along the area of the biaxial blend that had more of the corncob ash, and so suggested that the corncob ash is not as siliceous as the rice husk ash. Therefore, the corncob ash may not be a direct substitute of the rice husk ash.

Conclusion

Our experience shows that Chun glazes remain rather fugitive, given such variables as firing cycle and ash materials. For example, the same recipe (W) that turned out a dense Chun in one test had earlier produced yellow Chuns with very weak blue color.

Obviously, a number of factors are critical in the making of Chun glazes: prolonged firing at high temperature would cause the silica and phosphorus to go into solution, turning the glaze transparent. Another effect of prolonged firing was noted when recipe AF, which had produced a dense white emulsion in one test, turned out yellow in a firing that took a long time as a result of power failure.

The thickness of application is also of great importance, as some glazes that were dipped thin tended to go transparent even when fired correctly.

While the Chun effect may not be uncommon with glazes that have been fluxed with ash, the fact is that the temperature range at which the colloidal particles exist is small and only under controlled circumstances do they survive to give a blue, usually in places where the glaze is thick. Whatever approach is adopted, the making of these Chun glazes demands quite a lot of the potter's patience. ▲

Awka Oil-Spot Glaze

by Emman Okunna

Wheel-thrown pitcher and mugs, 8 inches in height, with Awka Oil-Spot Glaze, fired to Cone 10 in oxidation.

At Azikiwe University in Awka, Nigeria, we had been fascinated by the rich black glazes of China's Song dynasty, classified by the Japanese as temmokus, which have the remarkable property of breaking to rust. We had therefore been running several tests, trying to recreate this glaze with local materials, when a propitious accident occurred.

In the course of conducting these tests, a blending of two recipes, one lime (AL) and the other alkaline (AR), produced some good plain black glazes, but these blends showed a strong tendency to run, indicating that AR was not sufficiently alkaline. We, therefore, decided to blend AR with another recipe (BB) with slightly increased alkalinity and reduced ash (lime) content:

	AR	BB
Sawdust Ash (washed)	15%	10%
Talc	5	5
Emene Clay (sieved)	25	25
Granite	55	60
	100%	100%
Add: Red Iron Oxide	4%	4%

All tests were fired to 1280°C (2336°F) in both oxidation and reduction.

The result was most unexpected. While we were looking forward to a rich plain black glaze, without a tendency to run, the blends that were higher in BB produced a rich black glaze with rust spots. The blends that were higher in AR came out plain black, evidently the result of the lime taking the iron oxide completely into solution. When the lime content was reduced, it became possible to have some free iron, which produced the spots. Our lime glazes would not produce any blacks spotted with rust despite increased quantities of iron oxide.

Of immense interest to us was Nigel Wood's emphasis, in his book *Oriental Glazes*, on those materials (dolomite and whiting) that he said would aid the oil-spot effect, as his recipe was quite different than the one we had evolved:

Nigel Wood Oil-Spot Glaze
(Cone 9)

Dolomite	15.60%
Wollastonite	1.83
Potash Feldspar	18.35
BB Ball Clay	64.22
	100.00%
Add: Red Iron Oxide	4.59%

Awka Oil-Spot (BB) Glaze
(Cone 9)

Sawdust Ash (washed)	10%
Talc	5
Emene Clay (sieved)	25
Granite	60
	100%
Add: Red Iron Oxide	4%

Perhaps significant, too, are the results we obtained using iron spangles (Fe_3O_4) in some batches. These produced rather metallic-looking glazes, spotted with rust on a brown background. The spots were more frequent, but smaller in size. The batches that were mixed with red iron oxide (Fe_2O_3) produced really black glazes with large reddish brown spots. These results were the same in both oxidation and reduction firings.

The effect produced by the iron spangles is perhaps the result of the hard and dense nature of the material, which makes it resistant to chemical breakdown, as noted by Frank and Janet Hamer in *The Potter's Dictionary of Materials and Techniques.*

When the recipe with iron spangles was applied on a body dipped in white slip, the metallic luster was completely removed, producing a soft oil-spot glaze on a brown background. The white slip may have "starved" the glaze of the extra iron it had hitherto taken on from the clay body.

The thickness of application was also of great importance. There were usually more spots where the glaze was thick.

While all the details of our tests have not been included, it is apparent that the oil-spot effect can be obtained from a variety of materials, and is not dependent on the availability of particular commercial ingredients. What is important is to introduce a good balance of appropriate materials (flux, alumina and silica) to make the glaze sufficiently alkaline. Whatever the final combination, the addition of iron oxide plays the strongest part; it brings about various spotting effects, depending on the type of iron oxide used. ▲

Tiles with oil-spot glaze in which different types of iron oxide were used, fired to Cone 10 in oxidation, by Emman Okunna, Awka, Nigeria.

Electric Kiln Copper Reds

by Robert S. Pearson and Beatrice I. Pearson

Bright red was achieved by mixing Base Glaze 1 with 2.8% Color Mixture 4.

Copper red glazes are formed by reducing copper compounds in the presence of tin. Most often the reducing conditions are obtained by controlling the gas-air ratio in a gas-fired kiln; however, a reducing environment can be obtained in an electric kiln by including a suitable reducing agent within the glaze mixture. In recent tests, we had some initial promising results with both powdered elemental silicon and powdered elemental boron, but neither of these is as widely available as silicon carbide, the reducing agent in all the Cone 5 glazes described here.

A large number of glaze compositions are suitable for producing copper reds. Nevertheless, there is a much larger number of glaze compositions that refuse to produce any hint of red; e.g., glazes containing too much clay gave no reds. In fact, a higher than normal silica content seemed necessary to give good reds.

The red color varies somewhat from glaze to glaze. Higher calcium glazes shift the red slightly toward blue; however, all of the reds described below come closer to true red than any glaze stain we have tried.

Just for the record, our kiln has a capacity of about 3 cubic feet, we raise the temperature at the rate of 350°F per hour and fire to 2220°F, at which point a self-supporting number 5 cone bends enough to touch its support.

The color-forming ingredients are added to the glaze batch as a single addition, which we call the color mixture: copper carbonate, tin oxide, silicon carbide and a diluent. The very small amounts of copper carbonate and silicon carbide used are difficult to weigh directly. For this reason, they are combined with the tin oxide and the diluent.

There is nothing special about either the tin oxide or the copper carbonate, but the silicon carbide requires some discussion. Silicon carbide is available in a wide range of particle sizes. The coarsest that seems to be suitable in glazes is sold as 3F or FFF. Our sample of 3F silicon carbide was obtained many years ago from a lapidary supply house. Examination under a hand lens shows that it consists of a wide range of particle sizes. Although 3F is nominally 320 mesh, it is clear that our sample contains considerable fine material, so we assume it is a mixture of all particle sizes that pass through a 320-mesh sieve.

A purple glaze with blue flecks was obtained by adding 3.5% Color Mixture 3 to Copper Red Base Glaze 7.

Trials with 320-mesh silicon carbide have not produced useful glazes.

A more uniform fine-grained material sold as 1000-mesh silicon carbide is available from some laboratory supply houses and lapidary suppliers. One source is Ward's Earth Science, Post Office Box 92912, Rochester, New York 14692-9012.

The very finest particle size material is sold as ultrafine silicon carbide with individual particles in the micron-size range. It comes in three size grades: number 5, 10 and 15, with 15 being the finest. We used ultrafine 10 in all our tests. Unfortunately, at this time we are unaware of a retail source for this material. The bulk dealer is H. C. Starck Company, 45 Industrial Place, Newton, Massachusetts 02161. Although their minimum order is more than any one potter is likely to want, a group of potters might consider the amount affordable. Also, if there is sufficient interest in this material, a ceramics supply house may be persuaded to add this product to its inventory. This ultrafine-particle-size silicon carbide gave us the brightest and clearest reds; however, any very-fine-grained silicon carbide should be worth a try.

The diluent, which is added simply to make weighing easier, is always one of the major glaze ingredients. We used either flint or Custer feldspar as diluent.

The color mixtures most often have the following compositions:

Color Mixture 1

Copper Carbonate	7.6%
Tin Oxide	50.0
Flint	22.4
3F Silicon Carbide	20.0
	100.0%

Color Mixture 2

Copper Carbonate	4%
Tin Oxide	50
Custer Feldspar	36
1000-Grit Silicon Carbide	10
	100%

Color Mixture 3

Copper Carbonate	6%
Tin Oxide	47
Flint	40
Ultrafine 10 Silicon Carbide	7
	100%

Color Mixture 4

Copper Carbonate	5.0%
Tin Oxide	50.0
Custer Feldspar	39.4
Ultrafine 10 Silicon Carbide	5.6
	100.0%

Color Mixture 5

Copper Carbonate	3.6%
Tin Oxide	50.0
Flint	42.4
Ultrafine 10 Silicon Carbide	4.0
	100.0%

The brightest reds are usually formed with color mixtures having the least possible amount of copper. For best results, all ingredients in the color mixtures should be thoroughly blended, then ground together.

A white surface on the clay body is also essential for bright red glazes. Nonwhite firing clays can be coated with white slip, such as this one from Bernard Leach:

White Slip

Kona F-4 Feldspar	20%
Edgar Plastic Kaolin	60
Kentucky Ball Clay (OM 4)	20
	100%

The frits used in all of the glazes described below were obtained from General Color and Chemical, Post Office Box 7, Minerva, Ohio 44657. One glaze that gives good reds is:

Copper Red Base Glaze 1
(Cone 5)

Custer Feldspar	32%
Strontium Carbonate	4
Frit 146	32
Flint	32
	100%
Add: Macaloid	1%

A fair red can be obtained by adding 2.8% Color Mixture 2. A 2.0% addition of Mixture 4 or a 4.5% addition of Mixture 5 gives very nice reds.

Copper Red Base Glaze 2
(Cone 5)

Custer Feldspar	30%
Frit 146	30
Frit 154	10
Flint	30
	100%
Add: Macaloid	1%

Either a 1.5% addition of Color Mixture 4 or a 2.5% addition of Mixture 5 gives good reds.

Copper Red Base Glaze 3
(Cone 5)

Custer Feldspar	32%
Frit 146	32
Frit 156	4
Flint	32
	100%
Add: Macaloid	1%

A fairly good red is obtained with 3% Color Mixture 2, and a nice bright red is obtained with either 2% Mixture 4 or 3.2% Mixture 5.

Part of the frit is replaced by Borax in Copper Red Base Glaze 4
(Cone 5)

Borax (10 mol)	10%
Strontium Carbonate	4
Custer Feldspar	33
Frit 146	23
Flint	30
	100%
Add: Macaloid	1%

A very bright red is obtained with the addition of 2.5% Color Mixture 4 to Base Glaze 4.

Base Glaze 5 has a number of minor variations that are all useful. A nice red is obtained with 3% Color Mixture 4:

Copper Red Base Glaze 4
(Cone 5)

Borax (10 mol)	10%
Strontium Carbonate	4
Custer Feldspar	33
Frit 146	23
Flint	30
	100%
Add: Macaloid	1%

A very bright red is obtained with the addition of 2.5% Color Mixture 4 to Base Glaze 4.

Base Glaze 5 has a number of minor variations that are all useful. A nice red is obtained with 3% Color Mixture 4:

Copper Red Base Glaze 5A
(Cone 5)

Whiting	7%
Custer Feldspar	31
Frit 146	31
Flint	31
	100%
Add: Macaloid	1%

Copper Red Base Glaze 5B
(Cone 5)

Whiting	6%
Custer Feldspar	34
Frit 146	30
Flint	30
	100%
Add: Macaloid	1%

Using 3.5% Color Mixture 4 in variation 5B gives a nice red that can be applied thinly.

A third variation (5C) gives a blued red with the color mixtures:

Copper Red Base Glaze 5C
(Cone 5)

Whiting	10%
Custer Feldspar	30
Frit 146	32
Flint	28
	100%
Add: Macaloid	1%

Copper Red Base Glaze 6
(Cone 5)

Strontium Carbonate	4%
Custer Feldspar	42
Frit 111	28
Flint	26
	100%
Add: Macaloid	1%

Base Glaze 6 gives a cranberry red when combined with 3.0% Color Mixture 3.

As noted earlier, clay content tended to impede red color; however, these glazes made without clay minerals proved to be difficult to apply. We found the use of Macaloid at the rate of about 1% to be helpful; and instead of mixing the dry glaze with plain water, we use a 0.75% suspension of CMC gum (1 milliliter of gum suspension for each gram of glaze). The resulting viscous glaze suspensions are slower to dry, but give fairly uniform coverage.

When the glaze is applied with a brush, it is often necessary to apply four glaze coats to obtain the desired thickness and uniformity. Because of the added moisture from four brushed-on coats, we found it necessary to glaze the outside of a bowl first, allowing it to dry for a few hours or overnight before glazing the inside.

Reglazing poorly covered pots has not been successful. A second firing destroys part of the red color developed in the original firing.

A few substances were found to cause major color changes when added to some of the glazes. Rutile and bone ash were especially active. Those susceptible glazes contained moderate amounts of calcium either as whiting or as frit 111.

Rutile was very effective at shifting glaze color. A gradual increase in rutile content shifts the glaze color from red through various purples to blue. Continued addition of rutile causes the blue to fade. An almost endless range of red-purple, purple, blue-purple colors can be produced by varying the color mixture/rutile ratio and by varying the total amount of color-forming materials. Of course, changing from one color mixture to another will also have an effect on the final color. To a lesser extent, bone ash also shifts the glaze color toward purple.

The high whiting variation of Base Glaze 5 (5C) shows the color shift. When 3% Color Mixture 3 is added along with 3% rutile, an intense blue results; while the addition of 2.5% Color Mixture 1 along with 3% rutile gives a blue glaze with a red pattern.

Base Glaze 6 yields a wide range of colors with rutile. Adding 5% Color Mixture 4 along with 2% rutile gives a purple with pink; 3% Mixture 3 along with 2% rutile yields a blue-purple; and a thinner application of the glaze containing 4% Mixture 4 and 2% of rutile results in red-purple. A 1.6% addition of Mixture 1 gives a red-purple without adding any rutile.

Base Glaze 7, which is high in calcium, also yields purple instead of red without an addition of rutile or bone ash with some color mixtures:

Copper Red Base Glaze 7
(Cone 5)

Strontium Carbonate	4%
Custer Feldspar	28
Frit 111	32
Flint	36
	100%
Add: Macaloid	1%

When 3.5% Color Mixture 3 is added to Base Glaze 7, the fired result is a red-purple with small blue flecks. Base Glaze 7 also gives a blue with pink flecks when combined with 2.5% Mixture 1 and 3% rutile.

Actually, if the silicon carbide is left out of the glaze, a very pale blue results. So little copper is present in the best red glazes that when no reduction takes place the glaze is almost colorless. Slightly increasing both the copper carbonate and tin oxide over what is needed for a red glaze gives a light blue glaze:

Base Glaze 8
(Cone 5)

Custer Feldspar	40.0%
Frit 111	26.0
Flint	34.0
	100.0%
Add: Tin Oxide	2.5%
Copper Carbonate	0.6%
Macaloid	1.0%

Adding 5% praseodymium yellow stain shifts the color to light green.

With the proper white clay surface (nonwhite clays can be coated with white slip) and sufficient care in the glaze application, satisfactory and reproducible copper reds, along with several purple to blue color variations, can be obtained in an electric kiln. ▲

More Electric Kiln Copper Reds

by Robert S. Pearson and Beatrice I. Pearson

In the November 1997 issue of *Ceramics Monthly*, we described a series of copper red glazes produced in an electric kiln with local reduction from silicon carbide. These glazes worked well on a Cone 5 stoneware body, but crazed badly on a Cone 6 porcelain body. As noted in the earlier article, stoneware bodies, even those described as "white" firing, are never white enough to produce the best reds, and it is often necessary to coat the body with a white slip for optimum results. The use of a porcelain body avoids this problem. The glazes listed below were developed to fit the Cone 6 porcelain body.

Copper Red Base Glaze 9
(Cone 6)

Gerstley Borate	6.0%
Custer Feldspar	22.0
Frit 146 (General Color)	26.0
Flint	46.0
	100.0%
Add: Color Mixture 4	3.1%

Copper Red Base Glaze 9 applied to wheel-thrown porcelain bisqueware (dipped in glaze batch for three seconds), fired to Cone 6 in an electric kiln.

The previously described glazes needed macaloid to aid application; however, the inclusion of Gerstley borate made it unnecessary to add macaloid to this glaze. As before, the frits were obtained from General Color and Chemical, Post Office Box 7, Minerva, Ohio 44657. Also, they now supply the Ultrafine 10 silicon carbide used to prepare color mixtures, as well as some prepared color mixtures.

The composition of the color mixture used here is:

Color Mixture 4

Copper Carbonate	5.0%
Tin Oxide	50.0
Custer Feldspar	39.4
Ultrafine 10 Silicon Carbide	5.6
	100.0%

Copper Red Base Glaze 9 gives a very nice red. When prepared with 100 milliliters water for each 100 grams glaze, it is suitable for dipping, provided the bisqueware (fired to Cone 04) is held in the glaze suspension for a few seconds. The intensity of the red color varies with the submergence time of the ware: we have found three to four seconds to be about right for porcelain bisqueware. Each user will need to run a few trials to establish optimum submergence time.

Copper Red Base Glaze 10
(Cone 6)

Frit 114 (General Color)	34.0%
Frit 154 (General Color)	6.0
Frit 156 (General Color)	4.0
Gerstley Borate	5.0
Flint	51.0
	100.0%
Add: Color Mixture 4	3.3%

A nice red. To use it as a dip glaze, we added 105 milliliters water to each 100 grams glaze and kept the ware in the glaze suspension for about two or three seconds.

Copper Red Base Glaze 11
(Cone 6)

Frit 114 (General Color)	38.0%
Frit 154 (General Color)	3.0
Gerstley Borate	9.0
Flint	50.0
	100.0%
Add: Color Mixture 4	3.1%

Base Glaze 11 yields a very good red. To use it as a dip glaze, we added 120 milliliters water to each 100 grams glaze and kept the ware submerged for about six seconds.

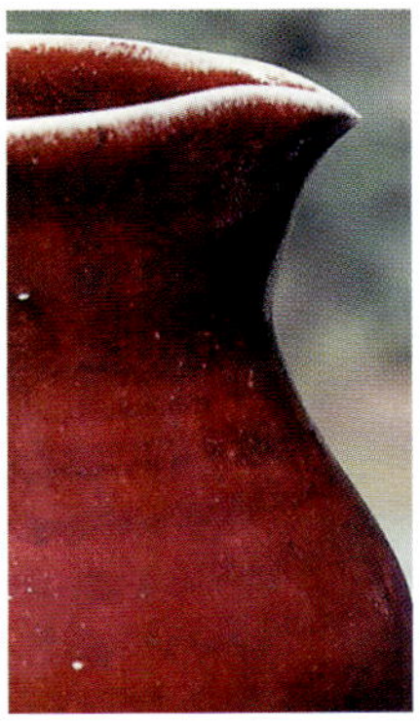

Cone 04 bisqued porcelain dipped in Copper Red Base Glaze 11 for six seconds, then fired to Cone 6 in an electric kiln; dipping time will vary according to clay body.

Copper Red Base Glaze 12
(Cone 6)

Frit 146 (General Color)	30.0%
Frit 154 (General Color)	10.0
Custer Feldspar	19.0
Whiting	2.0
Flint	39.0
	100.0%
Add: Macaloid	1.0%
Color Mixture 4	3.1%
Soda Ash	1.0%

When mixed with 100 milliliters of water for each 100 grams of glaze, a dip of approximately ten seconds gives a nice red.

Base Glaze 13
(Cone 6)

Frit 114 (General Color)	24.0%
Frit 154 (General Color)	16.0
Gerstley Borate	10.0
Flint	50.0
	100.0%
Add: Color Mixture 3	3.1%

Lavender-purple rather than red. Mix 110 milliliters water with each 100 grams glaze and dip for about three seconds.

Base Glaze 13 applied to wheel- thrown porcelain bisqueware, fired to Cone 6 in an electric kiln.

Color Mixture 3

Copper Carbonate	6%
Tin Oxide	47
Flint	40
Ultrafine 10 Silicon Carbide	7
	100%

Applied to stoneware, the lavender-purple is somewhat blued, but over stoneware coated with white slip is more red. Our stoneware and porcelain clays absorb glaze at different rates, so the stoneware requires a longer dip time, typically 10 or even 15 seconds, as compared to 3 to 6 seconds for porcelain. Clearly, a few time trials are needed to determine the best dipping time for any given clay. When using a brush, it is usually necessary to apply four coats to develop good color. ▲

Cone 5 Blue Glazes

by Dwain Naragon

It can be difficult to find glazes for mid-range temperatures that have character and a wide range of color and surface value. The following recipes, fired in oxidation at Cone 5, are just such glazes. These blues are determined by adding copper carbonate, cobalt carbonate and rutile to potassium-, sodium-, lithium- and barium-influenced glazes. They work well on stoneware and porcelain. *Given recent health concerns about the use of barium, NT 33 Blue Glaze should be reserved for use on nonfood surfaces*; or substitute 0.75 strontium carbonate for every 1.00 part barium carbonate.

Wolly Glaze Base
(Cone 5)

Wollastonite	35.35%
Kona (F-4) Feldspar	35.35
Ball Clay	29.30
	100.00%

For a dark navy, add 4.00% cobalt carbonate; for medium light navy, add 4.00% cobalt carbonate and 5.00% to 10.00% Zircopax.

Copper Blue Glaze
(Cone 5)

Whiting	21.98%
Potash Feldspar	63.74
Flint	14.28
	100.00%
Add: Zinc Oxide	10.99%
Copper Carbonate	1.10%

Bright Blue Glaze
(Cone 5)

Whiting	21.98%
Potash Feldspar	63.74
Flint	14.28
	100.00%
Add: Zinc Oxide	10.99%
Cobalt Carbonate	0.50%
Copper Carbonate	1.10%

Rutile Blue Glaze
(Cone 5)

Whiting	20.0%
Potash Feldspar	58.0
Barnard Slip	3.0
Kaolin	6.0
Flint	13.0
	100.0%
Add: Zinc Oxide	10.0%
Cobalt Carbonate	0.5%
Copper Carbonate	1.0%
Rutile	4.0%

Floating Blue Glaze
(Cone 5)

Gerstley Borate	27.0%
Nepheline Syenite	47.3
Kaolin	5.4
Flint	20.3
	100.0%
Add: Cobalt Carbonate	2.0%
Red Iron Oxide	2.0%
Rutile	4.0%

G184 Altered Base Glaze (Conrad)
(Cone 5)

Dolomite	6.25%
Gerstley Borate	13.54
Whiting	8.33
Kona (F-4) Feldspar	47.92
Kaolin	3.13
Flint	20.83
	100.00%

Color variations are achieved with the following additions:

Royal Blue:

Zircopax	10.42%
Cobalt Carbonate	1.56%

Slate Blue:

Zircopax	10.42%
Black Stain 6666 (Mason)	0.52%
Cobalt Carbonate	1.56%

Blue/White:

Zircopax	10.42%
Cobalt Carbonate	0.26%

Periwinkle:

Zircopax	10.42%
Cobalt Carbonate	0.52%
Copper Carbonate	1.04%

Cobalt Blue:

Cobalt Carbonate	1.56%

Cobalt Blue Spodumene Glaze
(Cone 5)

Dolomite	21.82%
Gerstley Borate	12.73
Spodumene	27.27
Potash Feldspar	19.09
Kaolin	8.18
Flint	10.91
	100.00%
Add: Zircopax	5.00–10.00%
Cobalt Carbonate	0.25–2.00%

For a matt glaze, decrease the amount of Gerstley borate.

N2 Satin Blue Glaze
(Cone 5)

Dolomite	3.19%
Whiting	4.26
Wollastonite	19.15
Kona (F-4) Feldspar	44.68
Ball Clay	14.89
Kaolin	3.19
Flint	10.64
	100.00%
Add: Zircopax	5.00–10.00%
Cobalt Carbonate	0.50–2.00%

NT33 Blue Glaze
(Cone 5)

Barium Carbonate	9.68%
Dolomite	4.30
Whiting	17.20
Potash Feldspar	35.48
Ball Clay	11.83
Flint	21.51
	100.00%
Add: Zinc Oxide	7.53%
Zircopax	5.00–12.00%
Cobalt Carbonate	0.25–2.00%

Crystalline Glazes: A Precise Method

By Bevan Norkin

In order to produce crystalline glazes you must first know what it is you are trying to create. Although the developed glaze is called crystalline, the structures that shimmer and float in the glaze are not single crystals-they are aggregates of microscopic crystals. These aggregates may be large or small and quite varied in appearance, depending on the conditions under which they were produced. In order to successfully and consistently develop these aggregates in the glaze, several variables must be identified and controlled.

Basically, this is a problem of precipitating crystals in a molten glass composition. And even though what we are describing has been labeled a glaze, it is really a glass composition that is almost completely free of alumina (Al_2O_3) and boric oxide (B_2O_3). These chemicals act as stiffeners, preventing the glass from flowing from the piece. They also inhibit the growth of the desired aggregate and can even prevent the precipitation of any crystals at all.

The technique of precipitating crystals may remain elusive unless you understand the nature of the precipitation product (the aggregate) in relation to the glass in which it is grown. The same molten glass medium will always yield the same aggregate, provided the aggregate is grown at the same temperature. The fundamental problem here is that glass is fluid when it melts, and as the glass flows down a vertical surface, the oxides that make up the glass composition cascade downward at different rates. This means that the same combination of oxides, whether fritted or not, will produce an infinite variety of molten compositions, which may vary radically in oxide content versus the percentage in the original composition. The aggregate grown will vary greatly in appearance; in some cases, the quantities of those chemicals needed for crystal precipitation are reduced to the point that nothing at all grows.

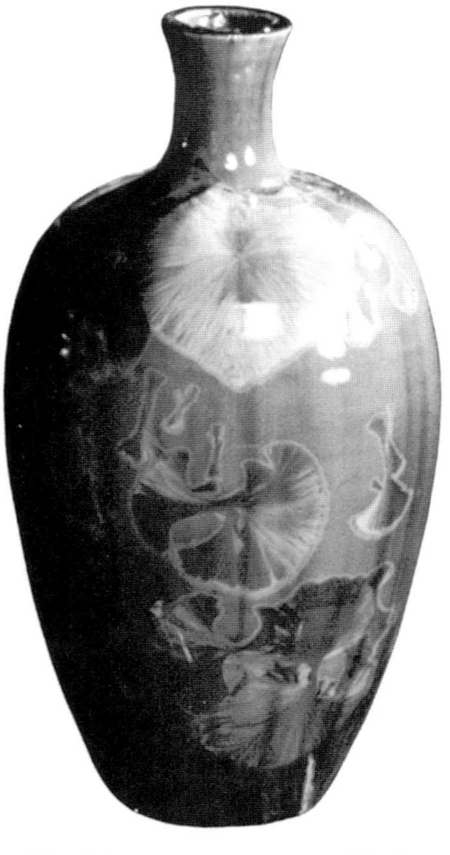

Wheel-thrown stoneware vase, 8 inches in height, with white-gold crystals on a gold background, fired to Cone 10 in oxidation.

An unusual feature of a glass composition that contains both zinc and silica is that ZnO and SiO_2 form two separate molten solutions that do not mix. It is apparent that the particle distribution in the initial raw glaze coat will influence how these two immiscible phases develop. It is absolutely essential that the initial glaze composition is properly suspended so that the same flow patterns can develop during the fluid phase. If you have a raw glaze that settles as you use it, no matter what the method of application, you will deposit a slightly different percentage oxide composition on different surfaces of the piece. Your piece will have, in effect, a number of different glazes deposited on the same surface. With a properly suspended glaze, you will find that the same result is obtained independent of the method of application (whether it is poured, dipped or brushed).

The method for producing a perfect suspension involves using 2% to 3% bentonite in combination with 1 gram of lithium carbonate (Li_2CO_3) per 500 grams dry weight of the oxides. Here the Li_2CO_3 acts as a deflocculating agent and the suspension produced is thixotropic, shifting from a gel to a liquid with gentle agitation. This suspension uses less water; the amount of water must be determined by trial and error. In general, I mix glazes in a food blender, adding water until the solution is very thick. At this point, the solution adheres to a brush without flowing, and bubbles rise out of the liquid and settle on the top without popping quickly. When allowed to sit for 10 to 15 minutes, then reagitated, the glop turns into a rich, creamy solution. If you have mixed it properly and allowed it to stand, the solution gels and no liquid accumulates at the top.

The behavior of the solution depends upon the ingredients selected. Western bentonite behaves as described. Other bentonites mix quickly and there is no waiting period or gloppy initial solution.

The glaze may be stored in an airtight container indefinitely with no settling. I have stored mixtures for as long as four months in Tupperware, and, upon opening, found no liquid on the top as well as no sediment at the bottom of the container.

Since the composition of the molten phase is so critical to the results obtained, you will find that the same raw glaze melted over different clay bodies also yields different results. My work has all been at Cone 9 to 12, so my technique may not apply at a lower temperature. Different porcelains may yield various results, such as a clear background with colored crystals or an opaque background with colored crystals. Such variation is related to the surface reaction between the clay body and the glaze. More brilliant results are obtained on less vitreous clay bodies. I have used both and somehow the results on semivitreous porcelain are crisper. The shape of the aggregate may change as well.

Firing Control

Forming the molten medium in which crystals are precipitated, mixing the glaze as a perfect suspension and selection of the clay body are the three variables that must be controlled, initially, to obtain repeatable results. However, you must also have a way to repeat the firing cycle. To this end, a programmable temperature controller is the best tool.

All the compositions that I have used have extremely broad ranges of temperature (from Cone 9 to Cone 12) during which the glass melts or become fluid. Of course, the aggregates that result are radically different. It should be understood that the glass is fluid as you approach the temperature selected as the melting point and also as the temperature descends from that point. I have never fired an entirely fritted composition, because I don't have the facilities to make one. My understanding is that fritted glass instantaneously melts at a specific temperature/time. But here we are using a frit in combination with raw oxides, so the range of melting is extended beyond a single point.

In firing this type of composition, it is necessary to plan a firing cycle. Any cycle will do-fast or slow. For the sake of economy I have used a fast cycle in this form:

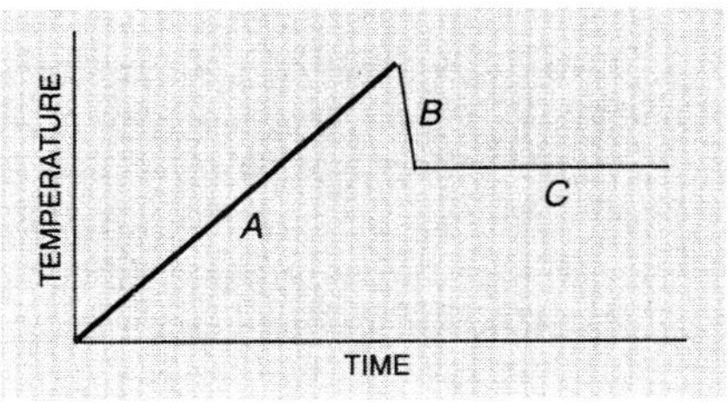

On the firing chart, A represents the increase in temperature to the melting point, while B marks the controlled descent to C, the temperature at which crystals precipitate out. The temperature-increase ramp may be set at any rate. My cycle begins with increases of 9°F/minute [5°C/minute] and gradually slows down to 4.1°F/minute [2.3°C/minute].

The descent should be rather fast. I have used two methods: continuous control from A to B to C, and broken control in which the kiln is shut off at the end of A, allowing the temperature to drop freely, then picking up control at about 54° to 108°F [30° to 60°C] below melting point for a controlled descent to holding temperature. The temperature is then held steady for 45 minutes to 6 hours. In general, I hold this temperature for 2-4 hours to allow aggregates to grow.

Using this method, crystals will grow at random with uniform distribution about the pot, provided there are no updrafts or crosscurrents in the kiln. These air patterns cause very small aggregates to form all over the piece or, particularly in the case of updrafts, cause crystals to form exclusively on the shoulders of the vase. If the updraft is confined to one side of the vase, bunching of aggregates will form on that side, while separate growth will occur on the other.

The various features of the aggregate are a product of heat treatment during their growth. At a constant temperature, the crystals will be deposited at a constant rate. Changing the growth rate by increasing and decreasing the holding temperature periodically will result in banding. If the temperature is allowed to oscillate about the set point, banded crystals will always occur.

After the kiln is turned off, the aggregate will continue to grow for a short time in the temperature zone in which the glass is still molten but not fluid. During this descent a fringe or halo will form about the crystal. The fringe develops in glaze compositions with a small percentage of coloring oxides, and a halo forms when a large percentage of coloring oxide is used (about 8%). Each of these shifts in configuration is produced by altering the holding temperature 36°F [20°C].

The temperature range in which crystal aggregates can be grown is extremely large and you will have to experiment with you own glaze. Various effects are produced simply by altering the holding temperature. At the top of the range, the aggregates formed look like small sticks with tufts along their borders or at either end. As you lower the holding temperature, they become (1) bowties lined up parallel to one another a perpendicular to the floor of the kiln, (2) fanlike plumes with variegated filaments, (3) fanlike plumes with parallel filaments, (4) circular with radiating parallel filaments and (5) splotchy centers with a filigree fringe.

If during the descent, the rate of temperature decline is held below 3.6°F/minute [2°C/minute] for compositions using nickel oxide, secondary aggregates that are golden and square may form, and two are sometimes connected by a bar. Then, generally, the Zn/Si aggregates form beneath them.

After the holding period is over and glazes with high concentrations of copper are left to cool, the lid of the kiln may be cracked slightly for 10 to 15 seconds, then shut. The draft that enters the kiln will cause secondary crystals to precipitate out in the form of black speckles at a temperature of 1292°F [700°C]. Holding the lid open several times will almost completely fill in the background with contiguous black dots, while holding the lid open too long will result in dunting of the ware.

Glaze Composition

In general, my method for experimenting with crystalline glaze compositions is to control the clay body, control the materials, use the same cycle over and over, and to alter the percentage composition of my recipe. Once that yields a result, I then adjust the holding temperature.

Most materials can be used directly out of the bag, except for Ferro frit 3110. Although this product is labeled 200 mesh, it turns out that this rating is for large quantities only. In order to obtain repeatable results, I have to pass it through a 200-mesh sieve. The lawn (mesh) of the screen must be made of stainless steel. It can be ordered in two configurations: open or standard. The open screen uses a smaller diameter wire and will sift the frit more quickly. It is important to dry sift this frit separately from the other ingredients instead of wet screening the entire mixture, because of the possibility of eliminating different amounts of other coarse materials from batch to batch. Sifting Ferro frit 3110 was something I discovered after all of the other variables had been identified; it was the key to obtaining repeatable results.

The zinc oxide also has a significant effect on results. I use 99.9% pure ZnO; the main difference from one manufacturer to another is the particle size. French-process calcined ZnO has a particle size of approximately 1 micron, and American-process calcined zinc has a diameter of approximately 0.19 microns. The difference in its use is the ease of mixing in suspension. The French process mixes much more easily, while the American process produces a more rubbery looking solution. Both work and produce beautiful results. But I found that some ZnO manufacturers change their method of producing ZnO and this would pull the rug right out from under my results. Finally, I found that Eagle Zinc (a product of Fusion Ceramics in Carrollton, Ohio) is equivalent to the French-process ZnO that I was using, plus the company guarantees it will not change the process, which has been in use for 63 years. If you are looking for consistency in results, this zinc is worth the difference in price. When I complained to my original suppliers, they simply replied I was not the only one who had complained. So zinc consistency is really an important consideration.

Another way of altering results and achieving subtle coloration in the crystal and in the background is to layer color variations of the same glaze recipe. For example, consider combinations of variations made from the following base recipe:

Crystalline Glaze Base
(Cone 10)

LithiumCarbonate	0.18%
Zinc Oxide	28.77
Frit 3110 (Ferro)	43.59
Bentonite	2.18
Flint	25.28
	100.00%

Variation I: add 1.05% red iron oxide to produce white crystals on a pale yellow background.

Variation II: add 1.05% copper carbonate to produce green crystals on a pale green background.

Variation III: add 1.22% manganese dioxide to produce mauve/pink crystals on a mauve background.

Variation IV: add 0.23% cobalt carbonate and 0.70% manganese dioxide to produce blue crystals on a gray background.

Variation V: add 2.62% green iron oxide and 2.62% red iron oxide to produce royal blue crystals on orange to red/brown background.

Variation VI: add 1.22% manganese dioxide to produce pink crystals on a mauve background.

Variation III fired over II produces green crystals on a pink and green background; II fired over III produces whitish green crystals over an oyster gray-green background, etc. However, Variation I fired over VI over IV results in VI melting out and having no effect on the color.

Layering Variation I over IV results in pale blue crystals on a hazy, pale gray-green background; however, developing yet another variation by using the colorants from both I and IV produces blue crystals on a definite green background.

To adjust the melting point of the crystals using Variation I, increase or decrease zinc oxide and flint each by the same amount. An increase results in a more refractory composition, whereas a decrease lowers the melting point.

Copper oxide is a very striking colorant when used in high concentrations. Adding 0.1% copper oxide to the base glaze results in a green-black crystal on an opaque or a translucent dark green background. This is the composition that, while cooling will produce black splotches with silver borders in the background, if the lid is cracked around 1290°F [700°C].

Firing Support

Although some potters stilt, then grind the bases of their crystalline-glazed pieces, I find this method very cumbersome. Instead, my pots have thick foot rings that fit firebrick pedestals cut slightly smaller than the ring's circumference. After having been cut to size, the pedestals are dipped into a mixture of alumina oxide and water. The Al_2O_3 coating prevents the corrosive glaze that flows off the piece from melting the brick. The pot and pedestal are placed on tope of another brick (also coated with Al_2O_3) and the whole assembly is placed on the bed of Al_2O_3 at the kiln's bottom.

The thick ring on the bottom of the pot allows the pot to shrink on the brick during firing without falling over, and helps keep glaze from running underneath. Pieces fired to vitrification generally shear off the brick on their own; however, most pots have to have the brick chipped off, then the bottom must be ground smooth. For grinding, I use a Black & Decker angle grinder with 16-grit silicon carbide sanding disks produced by 3M Corporation for floor stripping.

The final step is to polish the feet. I had a local steel fabricator who had a laser saw cut a 1/8-inch piece of iron in a circle to fit a bat. The iron disk is fastened to the bat with a stainless-steel hose clamp. I then use silicon carbide grit (220 mesh) to polish the base perfectly flat, just like a glass blower polishes work on a stainless-steel wheel. ▲

No More Gerstley Borate

by Jeff Zamek

Gerstley borate, a popular glaze ingredient in North America, is no more. The mine founded near the small town of Shoshone, California, in 1923 by James Gerstley (hence the name Gerstley borate) and subsequently sold to U.S. Borax, Inc., has been closed.

The company's decision was based on environmental and safety factors that made continued operation potentially dangerous for the miners. Besides, less than 1000 tons were being mined annually, a small amount when compared to the million tons per year of other borate minerals produced by the company. From a business standpoint, it was not worth the cost of bringing the mine up to current safety standards.

The demise of Gerstley borate illustrates the fact that any of the commercial materials used by potters could eventually be discontinued. The reasons behind the closing of a mine are usually based on economic factors, not on the exhaustion of deposit. Potters in North America will remember Albany slip, once popular in many glaze recipes. There are still vast amounts of this high-iron, low-temperature clay in the ground; however, the mine site is now a parking lot for a shopping mall. As stockpiles of Albany have run out, potters have been forced to rely on substitutes.

In many instances, the mining of geographically plentiful materials is discontinued because the major industries that had required the material stopped placing orders. Potters buy less than 0.1% of the ceramic materials on the marketplace, so we shouldn't be surprised when what we consider to be major glaze or clay ingredients disappear due to economic considerations. The one constant in ceramics is that there will be changes, either because of variations within the raw material deposit or lower demand from industry.

How does a potter deal with the unstable nature of raw materials? Education is the most effective solution to this type of problem. Knowing the chemical and physical makeup of a raw material is critical in developing a workable substitution. Even when the raw material is still available, your studio supply may be exhausted. Rather than wait for a new bag of Custer feldspar, for example, it might be useful to know that G-200 feldspar can be used as a substitute in many glaze and clay body recipes. If your recipe requires Oxford or Kingman feldspar, Custer or G-200 will also be appropriate substitutions. Potters should know the components of every raw material so they can have the flexibility to make a change when it is required.

History of Gerstley Borate

Glaze recipes containing Gerstley borate were developed in the 1950s, and were handed down from teachers to students in college classes, and passed from potter to potter. Although it is soluble and has a variable chemical makeup that can periodically cause several glaze defects, such as pinholes, blisters, dry spots and crawling, Gerstley borate is appreciated because of its fluxing action and the multihued surface effects it produces in low-, mid- and high-temperature glazes. It is also a component in many raku glaze recipes, which require good color response and fast melting during short firing cycles.

The ability to impart mottled, varied opacity effects to the fired glaze surface is due to the two primary minerals found in Gerstley borate melting at different temperatures (ulexite melts at 1535°F and colemanite at 1652°F). This is especially welcome in electric-kiln-fired glazes where the static oxidation atmosphere can produce flat, lifeless surfaces.

What to Do

Now that Gerstley borate is no longer being mined, potters will need to explore various replacement options. Of course, buying large quantities of the available material and using it sparingly is possible; however, the original problem of replacement will eventually have to be faced. A number of solutions are possible, but the degree of success depends on the specific glaze recipe.

In glazes containing 5% or less Gerstley borate, it might be possible to totally remove it from the recipe without changing the surface texture or fired color of the glaze. In these recipes, Gerstley borate is acting as a secondary flux, and the primary flux will take over the action of melting the glaze.

The mineral colemanite offers the best choice for a substitution. It has a more consistent chemical composition than Gerstley borate and offers less contamination from "tramp" materials. Interestingly, many current glaze recipes now using Gerstley borate were originally developed for colemanite. It was only when the supply of colemanite was cut off to ceramics industries that Gerstley borate was used as a substitute. Some potters even use colemanite and Gerstley borate interchangeably, thinking they are the same raw material.

The two minerals are chemically different, colemanite having a set formula of ($Ca_2\ B_6\ O_{11}\ 5\ H_2O$) as compared with Gerstley borate's combination of ulexite ($Na_2O\ 2CaO\ 5B_2O_3\ 16\ H_2O$) with small amounts of colemanite ($2CaO\ 3B_2O_3\ 5H_2O$), probertite ($Na_2O\ 2CaO\ 5B_2O_3\ 10\ H_2O$) and gangue, a bentonitic clay or tramp material often found as small black/brown specks. However, on a practical level, Gerstley borate and colemanite are interchangeable as ingredients in many glaze recipes.

Colemanite, while not readily available at present, does appear sporadically on the market. One source in Argentina exports colemanite to the United States, but only in a 100-mesh size, which would be too coarse for a Gerstley borate substitute. The company can deliver 200-mesh colemanite, but requires a minimum order of 20 metric tons or 44,092 pounds. A major importer of raw materials is currently testing the 200-mesh colemanite and may negotiate an exclusive agreement should it prove suitable for use in potters' glazes.

In glaze recipes containing more than 5% Gerstley borate (which do not require its color and surface mottling effects), a substitution flux can yield the correct amount of melting. In the Cone 06 (1830°F) to Cone 04 (1940°F)[1] temperature range, various commercial frits, such as Ferro frits 3195, 3134 or 3269, can be tested as a one-for-one substitution for Gerstley borate.

Some potters avoid the use of frits because of their tendency to sink like a stone to the bottom of a glaze bucket, but that can be overcome with suspension agents, such as CMC or Veegum T. The higher cost of frit as compared to Gerstley borate should not be considered a factor either. The real cost of working in ceramics is your labor, not raw materials.

Apply the test glazes to vertical tiles, leaving about 1 inch unglazed on the bottom of the tiles and placing them on an old kiln shelf. (This is standard procedure when testing any new glaze.)

In glaze recipes fired above Cone 6 (2232°F), Ferro frit 3195 or sodium-based feldspars, such as nepheline syenite and Kona F-4 feldspar, can be substituted. Other possible substitutions would be potassium-based feldspars such as Custer or G-200. The specific substitute material would have to be determined by the firing temperature, texture and color of the original glaze.

The substitution of a frit or feldspar would eliminate two troublesome characteristics of Gerstley borate and to a lesser degree colemanite—namely, both raw materials are soluble and have a variable chemical consistency from batch to batch.

Start by testing with a one-for-one replacement in a recipe calculated to total 100, so possible later additions of 5 parts substitute material will yield 105 batch weights. If the result is underfired (looks too dry), try adding 5 parts additional substitute material to the glaze batch. However, if the glaze runs off the vertical test tile, it has too much flux and should be mixed with 5 parts less of the substitute material.

If the fired glaze color is not satisfactory, choose another substitute. In glaze recipes that rely on specific levels of B_2O_3 for a mottled surface or color effect, a direct substitute of colemanite, feldspars or frit for Gerstley borate might not produce an exact match in color or surface mottling. Further testing would be required to duplicate these effects.

In glazes that do not require a specific level of B_2O_3, the substitution of a frit or feldspar would eliminate two troublesome characteristics of Gerstley borate and to a lesser degree colemanite—namely, both raw materials are soluble and have a variable chemical consistency from batch to batch. Frits and feldspars are, for the most part, insoluble and have greater chemical consistency.

A major frit manufacturer has shown interest in developing a substitute for Gerstley borate that might be insoluble as well as consistent in chemical composition. Another advantage in using a frit would be that the high chemical water component of Gerstley borate would be absent due to reactions taking place in the frit manufacturing process. High levels of chemical water in raw materials can cause excessive glaze shrinkage during the first stages of firing. Such shrinkage is one cause of crawling (the defect causes the molten glaze to roll back on itself, exposing bare clay areas looking very much like water on a glass table). Of course, the project depends on the company's assessment of this frit's economic viability.

U.S. Borax, Inc., has also expressed interest in finding another calcium borate ore that will replace Gerstley borate. However, due to Gerstley borate's unique chemical composition, it remains to be seen if a large corporation will invest resources into a venture that has a low profit margin and a low demand. The economic fact is potters do not make up a significant market and therefore are not considered when a raw material is deleted from production.

Remember, the disappearance of Gerstley borate is not a new situation. Historically, it is just one of the many raw materials that is no longer available for use by potters. Since individuals or even large supply companies do not have the economic buying power to guarantee the future supply of any raw material, we must educate ourselves in the science and art of substitutions. The plain fact is that the availability of any given raw material is uncertain. What is certain is that Gerstley borate will not be the last material to become history. ▲

1. All temperature references are based on large Orton pyrometric cones heated at 270°F per hour.

Substitutions for Gerstley Borate

by Jeff Zamek

With the closing of the Gerstley borate mine by U.S. Borax, a popular but problematic glaze material is history. While some potters stockpiled hundreds of pounds of Gerstley, others have relied on their suppliers to come up with a workable replacement. Unfortunately, the rush to develop an alternative has led to differing degrees of success.

To evaluate the substitutes currently on the market, a series of tests based on a fairly common glaze recipe was conducted. Floating Blue, a Cone 6 oxidation glaze, is well known by potters; it contains a relatively high percentage of Gerstley borate and is valued for its mottled surface quality:

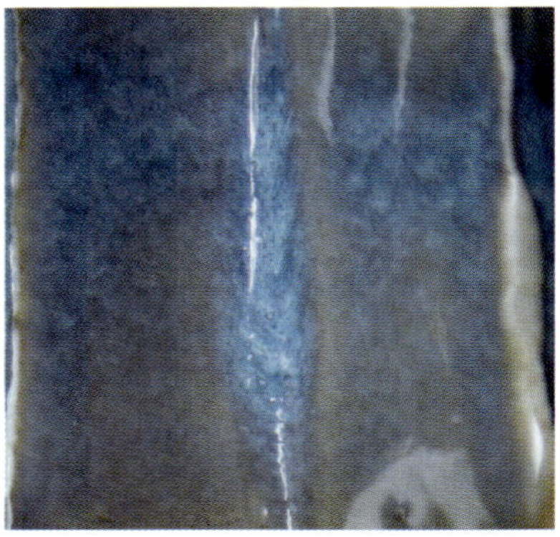

Original Floating Blue Glaze, mixed with Gerstley borate

Floating Blue Glaze
(Cone 6)

Gerstley Borate	26%
Nepheline Syenite	48
Edgar Plastic Kaolin (EPK)	6
Flint (325 mesh)	20
	100%
Add: Cobalt Oxide	1%
Light Rutile	4%
Spanish Red Iron Oxide	2%

As a control, the original recipe was mixed with Gerstley borate; then test batches with six different substitutes were mixed. Each was applied to tiles made from a commercial porcelain body (Portland Porcelain from Ceramic Supply of New York/New Jersey, 7 Route 46 West, Lodi, New Jersey 07644; 800-723-7264) that had been bisqued at Cone 06. Each tile was double dipped, resulting in a thin and a thick application. Three test tiles were made for each variation, and dispersed throughout the kiln to ensure accurate representation of each substitute.

In the control batch, Gerstley borate stayed in suspension, but also caused the glaze to become slightly thick during application. The fired glaze was a mottled, rich, milky blue.

Gerstley borate is not one mineral but a combination of minerals, and it often varies from one bag to the next, from one chemical analysis to the next. This variability has caused considerable difficulty in finding a substitute—in order to duplicate a material you first have to be able to accurately assess what the material consists of. The following is typical of a Gerstley analysis:

Gerstley Borate Chemical Analysis

B_2O_3	28.0
Na_2O	5.3
CaO	20.6
SiO_2	9.5
MgO	3.5
Al_2O_3	1.1
Fe_2O_3	0.3
LOI	25.0

So why did studio potters use Gerstley borate in the first place? The activity of natural borates, such as Gerstley, results from surface tension changes as calcium and sodium go into a melt in the presence of boron and silica, which are glass formers. During the firing process, a mechanical turbulence takes place, as calcium and sodium move back and forth to equalize their chemical potential, much like oil and water in the presence of certain detergents. The result of this interface turbulence is the variegation unique to Gerstley borate glazes. Gerstley borate can also cause crystallization in glazes and heighten color saturation.

At the same time, glaze batches containing Gerstley can become unstable in their dry or liquid form due to the borate's solubility. Moreover, many Gerstley borate glaze surfaces are "soft" or easily abraded and can leach.

So before rushing to replace Gerstley borate in glaze recipes, studio potters should stop to consider whether a substitute is even wanted. If the fired glaze color does not depend on a variegated, mottled effect, a frit can probably function as a substitute. If the glaze contains less than 5% (based on a 100% batch) Gerstley borate, it is not the primary flux, and the glaze will probably function correctly without it or by using an equal amount of frit (the particular frit will depend on the color, opacity and surface texture desired). For glazes that do not fit into either of these two categories, a substitute will be necessary.

Laguna Borate

One of the first substitutes on the market was Laguna borate (Laguna Clay Company, 14400 Lomitas Avenue, City of Industry, California 91746; 800-452-4862; www.lagunaclay.com). It is composed of a calcium borate in combination with raw materials and

Laguna borate test

chemical enhancement. In its description of the material, Laguna suggests several adjustments that can be made to a glaze batch, such as adding 2% bentonite for suspension or adding 0.5% CMC to prevent powdering.

Laguna Borate Chemical Analysis

Na_2O	3.97
K_2O	1.42
CaO	18.80
MgO	2.26
Al_2O_3	8.07
SiO_2	18.75
B_2O_3	27.52
TiO_2	0.07
Fe_2O_3	0.07
LOI	19.00

When substituted in the Floating Blue recipe, Laguna borate was easy to mix and afforded an even application. The fired glaze was dark blue, with minor pinholes. There was a slight milky opacity in some areas, but it did not duplicate the Gerstley borate effect in this type of glaze.

Murray's Borate

Kickwheel Pottery Supply (6477 Peachtree Industrial Boulevard, Atlanta, Georgia 30360; 770-986-0911; www.kickwheel.com) developed Murray's borate by combining a frit with raw materials.

Murray's Borate Chemical Analysis

SiO_2	29.25
Al_2O_3	0.50
CaO	26.21
K_2O	6.02
B_2O_3	38.03
Na_2O	6.02
LOI	5.0

When used in the Floating Blue recipe, Murray's borate did not settle and offered fairly easy application. The fired color and variegation were similar to that achieved with Gerstley borate, but the glaze tended to run slightly on vertical surfaces, exposing a thinner, darker color. Glaze running could easily be controlled by a slight addition of Edgar Plastic Kaolin to the recipe.

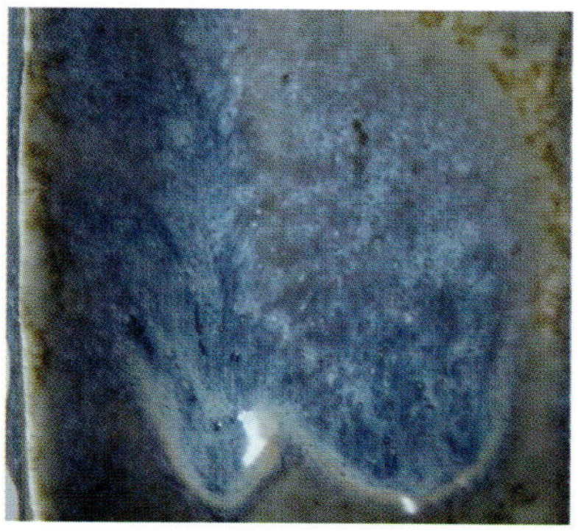

Murray's borate test

Cadycal 100

A calcium borate produced by Fort Cady Minerals Corporation (Post Office Box 100, Newberry Springs, California 92365; 760-257-4079; www.cadycal.com), Cadycal 100 is created by precipitating the borate from in situ mine leach liquor. The process uses diluted sulfuric acid that is driven into wells that are located approximately 50 miles from the U.S. Borax deposit where Gerstley borate was mined. When brought to the surface, the liquid is boric acid, which is reacted with lime, producing a pure calcium borate. The company lists several benefits, such as consistently high boron content, low loss on ignition, high calcium content and low impurities.

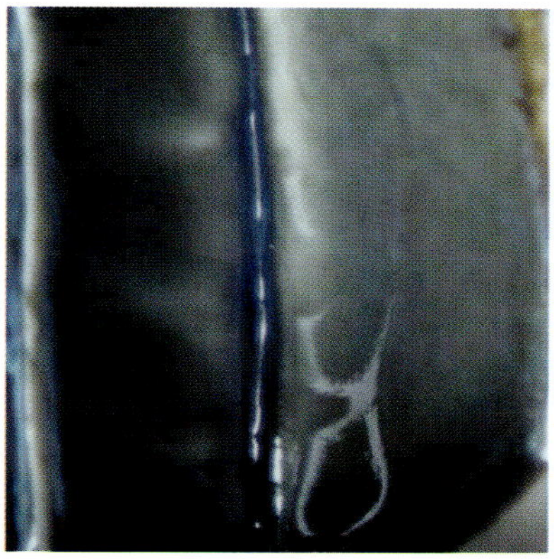

Cadycal 100 test

Cadycal 100 Chemical Analysis

B_2O_3	48.5
CL	0.10
SO_3	0.80
As_2O_3	1ppm
SiO_2	1.00
Fe_2O_3	0.08
CaO	27.00
Na_2O	0.10
Al_2O_3	0.20
K_2O	0.02
MgO	0.30
SrO	0.04
LOI	26.00

With Cadycal 100, the glaze cracked slightly when drying on the bisqued test tile. While the test batch did yield small areas of mottling, it produced a dark blue gloss glaze, which ran on vertical surfaces. Cadycal 100 is not recommended as a substitute for Gerstley borate in this type of glaze.

Gillespie Borate

A blended borate produced by Hammill & Gillespie (154 South Livingston Avenue, Post Office Box 104, Livingston, New Jersey 07039; 973-994-3650; www.hamgil.com), Gillespie borate contains a mixture of ulexite, various clays, alkaline earth carbonates and silicates. It has an identical match

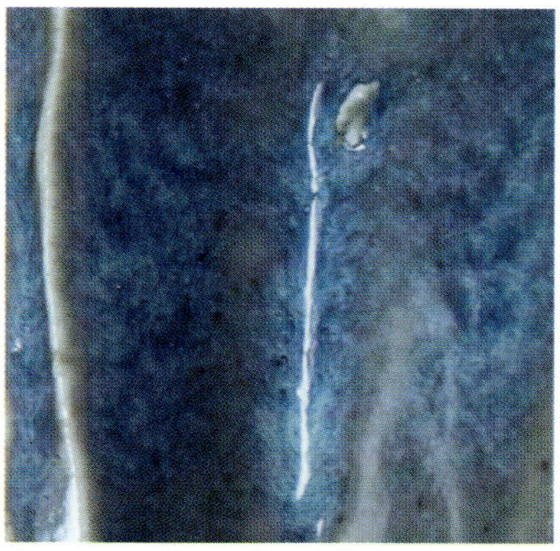

Gillespie borate test

in oxide content to Gerstley borate. Hammill & Gillespie claims it can replace Gerstley borate on a one for one basis, and it contains few impurities, which results in brighter glaze colors.

Gillespie Borate Chemical Analysis

SiO_2	11.80
Al_2O_3	1.70
B_2O_3	24.50
CaO	23.00
MgO	3.90
SrO	0.45
K_2O	0.01
Na_2O	3.77
LOI	30.90

Gillespie borate worked very well in the Floating Blue recipe, producing a successful match in color as well as the variegated effect. When mixed in the glaze batch, it did not settle and the glaze was easy to apply to the test tiles. Gillespie borate would be an acceptable substitute for Gerstley borate in this type of glaze.

Boraq

Boraq was formulated by Tony Hansen (IMC, 134 Upland Drive, Medicine Hat, Alberta T1A 3N7; 403-527-2826; www.gerstleyborate.com) from Cadycal, ulexite, feldspar and hectorite, all of which are higher quality materials than Gerstley borate, in that quality-control testing is done on every batch to ensure consistency.

Boraq Chemical Analysis

B_2O_3	34.5
Na_2O	3.4
CaO	17.8
MgO	2.2
K_2O	0.6
Fe_2O_3	0.2
SrO	0.3
Al_2O_3	2.5
SiO_2	12.5
LOI	26.0

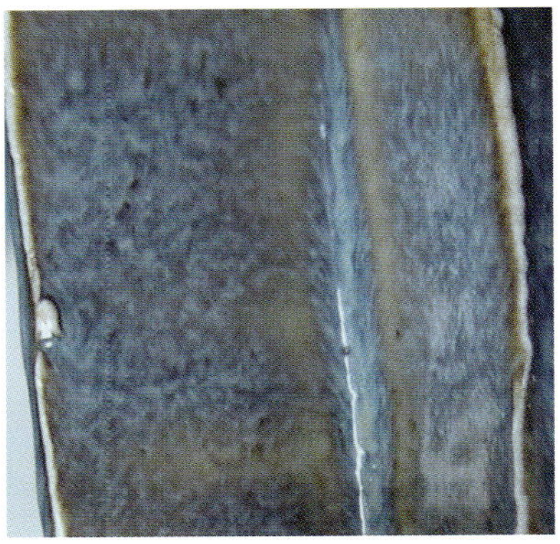

Boraq test

Boraq produced an effect similar to the glaze with Gerstley borate. The color and surface were exceptional. The batch mixed easily, remained in suspension and was trouble free in application. While there have been several revisions of Boraq since its inception, this sample worked well in this type of glaze.

Ferro Frit 3134

Produced by Ferro Corporation (4150 East 56th Street, Post Office Box 6550, Cleveland, Ohio 44101; 216-641-8580), frit 3134 is a combination of oxides that has been calcined and fast cooled, then ground to a powder.

Ferro Frit 3134 Chemical Analysis

Na_2O	10.2
CaO	20.1
B_2O_3	23.2
SiO_2	46.5

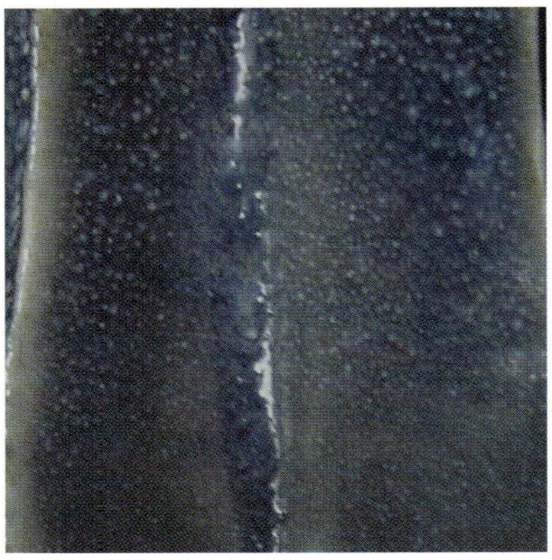

Ferro frit 3134 test

Because Ferro frit 3134 settled in the glaze batch, it would benefit from a suspension additive; however, the frit's major deficiency as a substitute in the Floating Blue recipe was its inability to produce a mottled blue surface. It fired to a dark glossy blue, with no mottling and slight pinholing.

Just how unique Gerstley borate is/was is now being underscored by how difficult it has been to find a simple one-for-one substitute in every glaze situation. Many manufacturers claim their product is chemically identical, but as we know, Gerstley borate had a somewhat ambiguous chemical makeup. With the manufacturers choosing a particular analysis for the basis of their substitute, the result can be slight differences in use.

For a low-quality, variable-content mineral, Gerstley borate gave many years of service to potters; however, several of the current substitutes are capable of producing better (i.e., more consistent) glazes due to purity and quality control.

Supply and demand have already determined the price structure of the Gerstley borate substitutes; however, purchasing decisions should not be based on price at any time. Choosing a material based on low price is a false economy. Time and labor factors are much more important.

The goal is to obtain materials that produce the desired effects on a consistent basis, while causing few or no defects. Using a low-cost material to save a few cents is not worth the risk if it does not perform well consistently. With this in mind, Gerstley borate substitutes should be chosen only on performance and reliability. ▲